Charles Karsner Mills

The Nursing and Care of the Nervous and the Insane

Charles Karsner Mills

The Nursing and Care of the Nervous and the Insane

ISBN/EAN: 9783337371340

Printed in Europe, USA, Canada, Australia, Japan

Cover: Foto ©berggeist007 / pixelio.de

More available books at **www.hansebooks.com**

PRACTICAL LESSONS IN NURSING.

THE

NURSING AND CARE

OF THE

NERVOUS AND THE INSANE.

BY

CHARLES K. MILLS, M.D.,

PROFESSOR OF DISEASES OF THE MIND AND NERVOUS SYSTEM IN THE PHILADELPHIA
POLYCLINIC AND COLLEGE FOR GRADUATES IN MEDICINE; NEUROLOGIST TO
THE PHILADELPHIA HOSPITAL; CONSULTING PHYSICIAN TO THE
INSANE DEPARTMENT OF THE PHILADELPHIA HOSPITAL;
LECTURER ON MENTAL DISEASES IN THE UNI-
VERSITY OF PENNSYLVANIA, ETC.

EDINBURGH:
YOUNG J. PENTLAND.
1887.

PREFACE.

The contents of this little book are the substance of a course of lectures delivered first at the *Training School for Nurses of the Philadelphia Hospital,* and subsequently at the *Woman's Hospital Training School for Nurses,* of Philadelphia. While books upon nursing are numerous, and a few valuable, no work devoted especially to the nervous and the insane had, so far as I know, appeared prior to the delivery of these lectures. Within a year or eighteen months, however, several books upon the care of the insane have appeared in England and in this country; but none on the nursing of patients suffering from either functional or organic nervous trouble, not forms of insanity. I have, therefore, felt that this little book might fill a want and serve a purpose. In no class of cases is it more important for a nurse, care-taker, or companion to have good principles of action and clear notions of practice, than among patients suffering from nervous or mental affections. Many of these unfortunates require prolonged and elaborate treatment, much of which necessarily must be carried out in the absence of the physician. I have frequently been asked by nurses where they could obtain, in compact form, some information as to the care of such patients, and also as to the use of massage, electricity, bathing, etc., by nurses. Many books upon electricity and massage have been published, but these are either too voluminous, containing too much scientific or theoretical matter, or they are not adapted to the comprehension and purposes of nurses. They are nearly all, especially the works upon electricity, books for physicians. Some books upon nursing contain suggestions or directions as to apoplexy, drunkenness, hysteria, epileptic seizures, etc.; but it will be found, on comparing these works with the present volume, that considerable additions have been made.

1909 CHESTNUT STREET, PHILADELPHIA.

CONTENTS.

CHAPTER I.

PAGE

Qualities and Qualifications of a Good Nurse for Nervous Patients—General Management of Hysteria—Hysterical Seizures—Epileptic Seizures—General Management of Epileptics—Forms of Insensibility—Cases of Chronic Organic Nervous Disease—Sleeplessness—Delirium—Diet—The Alcohol, Opium, or other Narcotic Habit 9

CHAPTER II.

Massage—Movements—Muscle-Beaters—Bathing — The Revulsor—Surface Thermometers 39

CHAPTER III.

Forms of Electricity—Toepler-Holtz Electrical Machine—Faradic Apparatus—Galvanic Apparatus—The Mechanism, Management, and Care of Faradic and Galvanic Apparatus—Hints, Cautions, and Contraindications in Using Electricity—The Milliampèremetre—Electrodes—Conducting Cords—Methods of Applying Electricity 63

CHAPTER IV.

The Nursing and Care of the Insane 106

LIST OF ILLUSTRATIONS.

FIG.		PAGE
1.	Muscle-Beaters of Ruebsam	56
2.	Toepler-Holtz Electrical Machine	66
3.	Flemming's Faradic Battery	67
4.	Dubois-Reymond Coil	68
5.	Faradic Apparatus for Office Table	70
6.	Portable Galvanic Battery	79
7.	Flemming's 30-Cell Combination Battery . .	80
8.	Barrett's Chloride of Silver Battery . .	82
9.	Diagram of Commutator	85
10.	Commutator Switch	86
11.	The Milliampèremetre	89
12.	Diagram showing Method of Using Milliampèremetre .	91
13.	Box of Electrodes	94
14.	Method of Holding Electrodes in one Hand . . .	95

THE NURSING AND CARE

OF

THE NERVOUS AND THE INSANE.

CHAPTER I.

Qualities and Qualifications of a Good Nurse for Nervous Patients—General Management of Hysteria—Hysterical Seizures—Epileptic Seizures—General Management of Epileptics—Forms of Insensibility—Cases of Chronic Organic Nervous Disease—Sleeplessness—Delirium—Diet—The Alcohol, Opium, or other Narcotic Habit.

THE nursing and care of the nervous and the insane call for special training in a high degree; and yet it is difficult to systematize and formally set forth instructions and regulations for such nursing. In fever, surgical, obstetrical, and other forms of nursing, much of the training has reference to instruments, dressings, temperature records, etc., to things which can be seen, handled, and illustrated. In the training of those destined to care for the nervous and the insane, a certain amount of instruction as to machinery and manual procedures must be given; but much of the teaching, from the very nature of the cases, must have reference to the conduct, habits, and characteristics both of nurses and patients.

In order, in the four chapters of the present book, to cover in a practical way my entire subject, I will not be able to give much time to theoretical considerations, and I believe that it is best that this course should be followed. While it is an advantage, and sometimes an important one, for non-medical attendants upon the sick to have some acquaintanceship with anatomy and physiology, or other branches of medical science, the amount of knowledge of this description absolutely requisite is limited.

The first chapter will be devoted to a consideration of some of the personal qualities and qualifications of nurses for nervous cases, and to certain miscellaneous matters in connection with nervous nursing; in the second, massage, movement, bathing, and certain instruments will be treated of; the third will discuss electricity; and the fourth will be on the nursing and care of the insane.

Diseases of the nervous system cover an enormous field. The last great work on these diseases, the fifth volume of "The System of Practical Medicine by American Authors," is a treatise of no less than thirteen hundred and twenty-six pages; but even this book, voluminous and exhaustive as it is, does not traverse the entire ground. Experience has taught me that while nurses who are called upon to care for the nervous and the insane may be needed to aid the doctor in almost any one of the long list of nervous and mental disorders, a rough attempt at arranging the different classes of cases in which they are most likely to be employed can be made, based upon the special qualifications that will be required, and the particular duties that will be imposed. First come the cases of functional nervous disorders, so called, under which head are included cases which furnish an important field for nurses thoroughly trained for their work,—patients suffering from hysteria in its manifold shapes; the neurasthenic or nervously exhausted; neurotic people in general, some of them on the border-land between sanity and insanity; the sleepless, the neuralgic, and the choreic. In the second place, we find patients suffering from some forms of insensibility and from certain acute organic affections,—cases of apoplexy, of uræmia, of sunstroke, of cerebro-spinal fever, of acute meningitis, tetanus, etc. Thirdly, there are certain cases of chronic, organic nervous disease, some curable and some incurable,—patients afflicted with meningitis cerebral or spinal, the victims of tumor of the brain or spinal cord, the hemiplegic, the paraplegic, or sclerotic, and the sufferers from atrophic or wasting diseases. Under a fourth head are included all cases of insanity.

At the outset let me speak briefly of some of the qualities and qualifications of a nurse for nervous patients. My first remarks will apply more particularly to those who will have under their care the sufferers from hysteria, neurasthenia, and other functional nervous disorders. The care of these patients is in some respects one of the highest forms of nursing. They sometimes undergo a method of treatment which involves, among other things, more or less complete seclusion from all but physicians and attendants; and, confined to their rooms for weeks or months, almost isolated from the rest of the world, their care demands high qualities and qualifications in the companion and care-taker.

Let me here quote the words of one[1] whose wide experience entitles him before all others to speak as to the trials and qualifications of those to whom fall the care of the nervously sick:

"A life spent beside such a sick-bed is indeed a test alike of character and health. It requires a strong body and a fortunate balance of moral and intellectual qualities to escape from being made morbid by constant contact with such suffering; and intensely sympathetic people are surely hurt by it, and themselves grow morbidly sensitive. Where the unhappy invalid becomes exquisitely ill-tempered under the long pangs of illness, the constant nurse must endure a thousand petty trials of temper, and must know when to yield and when to resist the tiny and numberless oppressions of her sick tyrant; but incessant battle with one's self is exhausting, and soon begins to show its results upon the healthiest nurse, cooped up in the sick-room. A pallid face, loss of energy, a certain passive obedience to routine duties are the sure consequences."

While the fundamental qualifications for a good nurse for nervous patients are those for a good nurse of any description, it is important that she should have a few special qualities, and should be free from a few particular

---

[1] 'Nurse and Patient," by S. Weir Mitchell, M.D., LL.D.

faults and foibles. It is in the highest degree important that she should possess the quality called tact, so often wanting in all walks of life. "Tact," says Dr. Anderson,[1] "is a quality not easily defined; but if we go back to the original meaning we can construe a definition upon it. It means, literally, touch,—the touch of skill and experience. But it has a wider significance; it includes the mental touch, something more complete than the other; not a touch merely, but a grasp, —the grasp of the situation, the comprehension of a difficulty, the grasping of it on all sides so that it disappears in your hands."

A domineering nurse will not succeed, while a decided one will have the best chance of success. Firmness should be combined with gentleness and fairness, and a genuine good temper is invaluable. Habits of close observation should be particularly cultivated. She should carefully note differences in the condition of the patient when the doctor is present and when he is absent; and any tendency to deceive or simulate symptoms should be quietly reported. She should be especially careful not to talk too much to the doctor about the patient before the latter. It is often best for her to retire from the room when the doctor is present, in order to give him an opportunity to talk to the patient in private. A good nurse will soon learn to discriminate as to when she should go and when she should stay, and when and how she should make her report.

It often falls upon the nurse to carry out plans and

---

[1] "Lectures on Medical Nursing," by J. Wallace Anderson, M.D.

invent expedients for amusing and occupying the patients. She should, therefore, have intelligence and education to be able to interest her patient by either conversation, reading, or other measures. She may fail in this respect by attempting to do too much, or by having an erroneous notion of her own powers. A good reader may interest patients of a certain class; a bad one may torture them. Judgment in carrying out directions as to amusement and occupation will call for exercise of high qualities. If reading is permitted, the right books should be chosen; if cards are resorted to, they should not be made a source of excitement instead of relief; whatever the occupation or amusement, it should be such as is adapted to both the physical and mental conditions of the patients. In many cases the nurse can persuade the patient into usefully occupying moments which would otherwise be passed in introspection.

"The nurse for these cases," says Mitchell,[1] "ought to be a young, active, quick-witted woman, capable of firmly but gently controlling her patient. She ought to be intelligent, able to interest her patient, to read aloud, and to write letters. . . . It is always to be borne in mind that most of these patients are over-refined, sensitive women, for whom the clumsiness or want of neatness or bad manners or immodesty of a nurse may be a sore and steadily-increasing trial. To be more or less isolated for two months, in a room with

---

[1] "Fat and Blood: An Essay on the Treatment of Certain Forms of Neurasthenia and Hysteria," by S. Weir Mitchell, M.D., LL.D.

one attendant, however good, is hard enough for any one to endure; and certain quite small faults or defects in a nurse may make her a serious impediment to the treatment, because no mere technical training will dispense in the nurse, any more than in the physician, with those finer natural qualifications which make their training available."

"What is required," says Playfair,[1] in a similar strain, "is a woman of kindly disposition and pleasant manners, and of sufficient intelligence and education not only to fully appreciate and second the object of the medical attendant, and to report to him the peculiarities of each individual case, but also to form an agreeable companion to the patient during her long seclusion."

A good nurse will always be loyal to the doctor; and indeed, to be loyal to him is always the best service to the patient. "Loyalty to the doctor," says Miss Weeks,[2] very forcibly, "includes encouragement of the patient's faith in him, so long as he is in charge of the case. The imagination is so largely active in disease that to infuse a doubt and distrust into the patient's mind is often to destroy all hope of doing him good. The nurse is a connecting link between doctor and patient, responsible to the one, and for the other, and can do much to promote good feeling between them." These words apply with particular force to nervous patients.

---

[1] "The Systematic Treatment of Nerve Prostration and Hysteria," by W. S. Playfair, M.D., F.R.C.P.

[2] "A Text-Book of Nursing." Compiled by Clara L. Weeks.

Having detailed some of the qualities and qualifications of a good nurse, let me picture some of the types of faulty nurses,—at least so far as the care of nervous patients is concerned; some of them faulty for any form of nursing. There is the nurse who talks too much. For some nervous patients, particularly for those who are just commencing a long course of treatment by rest and seclusion, too much talking on the part of the nurse is a serious drawback. With the constitutionally talkative nurse it is worse than useless to attempt to correct the habit. She will talk in season and out of season, and is usually unable to curb this propensity even in the presence of the physician; indeed, she sometimes seems to exhibit it as one of her greatest acquirements. On the other hand, I occasionally meet with a nurse who does not talk enough; or who never says anything when she does open her mouth; that is, an individual constitutionally wanting in conversational powers. Either of these extremes is bad. It is well that a nurse for nervous patients should talk sensibly, intelligently, and with judgment.

An abomination of abominations is the habit which some nurses have of informing the doctor as to the exact nature of the patient's disease; telling him that the case is one of hysteria; whispering an opinion as to its treatment in a mysterious manner; or boasting of some wonderful discovery with reference to the patient.

The too familiar nurse is also an abomination. Thrown closely in contact for weeks, and it may be for months, it is natural that a patient isolated from family and friends should become attached to the nurse who is kind and sympathetic, and at the same time firm and just. Even some affection may at times grow between the patient and the nurse; but it is the duty of the nurse, while neither morose nor distant, to preserve a quiet dignity and reserve of demeanor. The typical, natural, heaven-sent nurse will know just what to do in matters of this kind, but a few hints may be of value to some, who, although possessing many virtues, are not celestially endowed.

While it is of the utmost importance that all nurses should pay proper attention to personal appearance, the nurse who is vain and devotes too much of her time to self-adornment is especially obnoxious to nervous patients. She should not spend more time in preparing herself for exhibition than she does in attending to the interests of her patient.

The conceited nurse is a sore burden, but the nurse who is too humble is almost as bad. The nurse has her rights and a certain position to sustain. It is for her, like the soldier, to serve willingly and well, but not meanly and with mock humility.

The nurse who comes from a "good family" is sometimes a very good nurse, but may nevertheless be a sore trial to patient and physician. On the whole, it is not a bad thing to have had respectable ancestors; it is not to be placed to one's discredit if one's progenitors have been men and women of distinction; but it is better that this fact should appear in good breeding and bearing rather than that it should be reiterated into the ears of all comers.

The nurse who assures you that she is nursing simply because she loves the business is particularly unfitted for the care of nervous patients. Such an assertion generally means that the one making it either only half believes in, or is half ashamed of, her vocation. While nursing should not be regarded from a mercenary point of view, while those engaged in it should love it for its own sake as well as for what it brings, it is a business, and, at least, the effort should not be made to impress the outside world that an individual who is receiving fifteen dollars or more a week is actuated wholly by philanthropic motives. Such an effort will usually fail, particularly with those who settle the bills.

The nurse who quarrels with the servants is often a nuisance; although I must say on her behalf, that in my experience the latter are most frequently the aggressors. In the rest treatment, and in some other lines of nursing, the nurse is required to look after some special matters of diet, sometimes at unseasonable times, or to otherwise disturb the normal equilibrium of housekeeping. Servants naturally dislike those who, while not in the same position as the family, are on a plane higher than themselves. A nurse with tact and common sense who starts right in a family—except in special instances where an angel from heaven could not get along with the cook or chambermaid—will usually be able to at least so steer her course as not to come in actual violent collision with the servants; but she who attempts to put on airs or to lord it over the denizens of the kitchen will usually be badly routed in the end.

Before taking up special matters of nursing, let me say finally that I have noticed occasionally too great a disinclination on the part of the nurse who has been accustomed to a certain line of cases, to step out of that which she considers her particular sphere to take hold of patients of any other type. I regard it as a high quality in a nurse to be willing to turn to any form of nursing in her general line in an emergency. Nurses are not to blame for wishing cases of a class to which they have been accustomed, or for which they have tried to especially prepare themselves; but the practice of some of refusing, sometimes in an offensive way, to take anything unless they can get exactly what they wish, usually shows weakness of character, which will interfere with full success in their vocation.

I will now briefly consider the management of hysteria and epilepsy. First, before all, let me caution the nurse not to make the diagnosis of hysteria. Even physicians of large experience sometimes find it difficult to make a correct diagnosis in alleged hysteria, and it is a source of either amusement or amazement to me to see the self-confident nurse glibly pronounce a patient that she has seen perhaps once or twice as hysterical, or as an hysterical fraud. On the whole, I would advise the nurse never to use the word hysteria at all. To be hysterical is certainly not a disgrace; it is a condition into which the best of women, or, indeed, of men, may fall; but the idea which hysteria conveys to the common mind is too often that of shamming or fraud. Nine patients out of ten will feel that they are insulted if, in the midst of nervous suffering, they are called hysterical.

It is a foolish thing in a nurse who has charge of a patient suffering with hysteria to give way to every whim and fancy of the patient; and yet, on the other hand, she cannot by denunciation or harsh measures get rid of the inconvenience and care which such patients may give her. She should not do too much on her own responsibility in the way of severe treatment. The nurse may be called up three or four times during a night, or in some other way may be sorely tried, but all that she can do is to be watchful and careful, and if she finds that a patient is disposed to make things uncomfortable, by some mild but positive words or measures try to prevent her. She must not, as has been done in some cases, plunge the patient into a cold bath without directions from a physician. The one to use threats and severe measures, if these are resorted to at all, is the physician.

It may be a question sometimes whether a patient is suffering from a hysterical or from an epileptic fit, and this doubt is not always easy of solution, although usually the two affections can be separated. Hysterical seizures differ in character. Sometimes they are purposive or voluntary,—that is, they are completely under the control of the patient, by whom they are induced or feigned; at other times they are as completely beyond the control of the will as are the convulsions of epilepsy. The simulated or foolish fits of hysteria can generally be recognized. The screaming, shouting, and violent movements of the patient, at the same time that, perhaps, she can be discovered closely watching others, will put the physician or nurse on guard.

Some of the involuntary hysterical attacks, the paroxysms of hystero-epilepsy so called, closely resemble true epilepsy. These attacks are in some respects grave, although they are not either as incurable or as temporarily dangerous to life as genuine epilepsy. In the purposive or simulated hysterical fits consciousness is never lost. In the hystero-epileptic, or grave hysterical attack, however, it is not strictly true to say that loss of consciousness does not take place. It does occur, but there is a difference between the loss of consciousness in epilepsy and that in hystero-epilepsy. The question of consciousness is one of grade and degrees. A patient may be more or less unconscious. In epilepsy the unconsciousness is complete, profound, and persistent. In hystero-epilepsy consciousness is lost completely during only a short phase of the attack. The hysterical or hystero-epileptic patient rarely bites the tongue, while this accident commonly happens in epilepsy. I have observed in hystero-epileptic cases that whether the patient is or is not responsive to external impressions the countenance usually retains an unusual placidity. On one occasion, at the German Hospital in Philadelphia, I watched for more than two hours the frightful contortions of a typical hystero-epileptic; and, although during this long period she got into almost every imaginable grotesque and extraordinary position, during all the time her countenance remained as placid as a summer morning.

A nurse, or even a physician, sometimes makes a

mistake with reference to the diagnosis between an epileptic and an hystero-epileptic seizure through ignorance of the fact that sometimes the patient is the victim of both hysteria and epilepsy. A patient so afflicted is sometimes designated as a case of hystero-epilepsy, with separate crises. In the wards for diseases of the nervous system in the Philadelphia Hospital is a female patient who has had at long intervals well-marked epileptic seizures, accompanied by profound unconsciousness with biting of the tongue. In addition, however, at more frequent intervals she has had violent hysterical outbreaks. In such a case as this the nurse who is not well informed about such matters might readily make a mistake.

The nurse should have a clear idea of the characteristics of a genuine epileptic seizure. Usually such an attack comes on very suddenly, often preceded by a cry of pain or great fright. The patient falls suddenly with great force, sometimes in the most dangerous places. Convulsions then begin. Usually they are at first what is termed tonic,—that is, the limbs and body become rigid and immovable in certain positions. Then the so-called clonic movements occur,—that is, the patient is violently twisted, contorted, and tossed about. Frequently these movements begin in one limb or part and extend rapidly until all portions of the body are involved, and the convulsion becomes general. The face, at first pale, changes to red, gray, or purple, or becomes absolutely livid; or alternates between pallor, flushing, or lividity; the pupils are dilated and fixed; the eyes are frequently turned upwards; the patient froths at the mouth, ejecting often a mixture of saliva and blood. The tongue is bitten generally early in the seizure. The breathing is irregular and sometimes stertorous. The attacks vary in length, always seeming longer than they are, usually not lasting more than a few minutes at most. As the convulsive movements cease the patient sometimes becomes semi-conscious, but continues much dazed, or more commonly falls into a profound stupor. Sometimes urine is passed unconsciously, or the bowels are involuntarily evacuated. The pulse shows great variations.

When satisfied that a patient is suffering from an epileptic attack, it is important not to be too meddlesome. More harm may be done by unnecessary interference than by comparative neglect. It is important, however, that a person in an epileptic fit should be watched. If the clothing is binding anywhere, it should be loosened. The patient should be placed so as to have as little injury as possible result from the violent movement. It is better to have the head slightly elevated. If the room is close, fresh air should be allowed to enter it in abundance. It is not well to roughly seize and hold the patient, or to force open the clinched hands with great effort. It is not always best to make great efforts to open the mouth by means of a towel, or wedge, or bandage. The object of this would be to prevent the patient from biting the tongue, and as this commonly occurs early in the seizure, the mischief will usually have been done before the treatment can be applied. Here, as in so many other cases, no fixed rule applies; but good sense and quick wit come into play.

An epileptic attack cannot be aborted by brute force. Sometimes epileptics have, preceding the paroxysm, what is termed an aura,—that is, a peculiar feeling in some part of the body. It may be the sensation of a vapor passing up a limb, or a feeling of crawling, or a sharp pain. When it is known that such a sensation is felt at the extremity of a limb at each recurrence of the seizure, if the nurse is at hand the attack may sometimes be stopped by seizing and tightly holding the limb above the point where the aura begins. If with a patient known to be an epileptic, the nurse should see to it that the patient is never in any unnecessary danger; clothing should be worn a little loose, and the patient should be guarded from danger from fire or machinery.

Independently of the management of the paroxysm, a number of important practical points in connection with the general care of patients suffering from epilepsy are worthy of consideration. Before or after convulsion, and sometimes in the place of such an attack, epileptics sometimes pass into peculiar abnormal mental states. These may vary from a state in which the patient is simply dazed, confused, or dull, to an attack of acute mania. The nurse knowing this should be on the guard to prevent harm. Each patient should be carefully studied, and his or her peculiarities noted. Purposeless irritability and excitement, which are attributed sometimes by physicians and nurses to ill-temper, are really often the result of disease; and so likewise are states of suspicion and distrust. The diet in cases of epilepsy should be carefully attended to, as epileptics are likely to be large eaters. They seem sometimes to require large amounts of food; but they can easily eat too much. The amount and character of food are of course to be directed by the physician in charge, but to the nurse or attendant the carrying out of these directions must necessarily be intrusted. Observations made by some good authorities strongly favor a vegetable diet in epilepsy, and when such diet is ordered, the nurse should see that it is rigidly adhered to according to directions. Overloading of the stomach and prolonged constipation sometimes lead to the attacks; these should be prevented.

In a few cases it may be important for the nurse to be able to diagnosticate the cause or nature of insensibility, of other forms than that seen in epilepsy. While diagnosis is the duty of the physician, a little exact knowledge of a disease or condition may enable a nurse to so act in an emergency before the doctor arrives, or when he cannot come, as to give the patient a better chance of ultimate recovery, or possibly even to save life. The nurse should have some general idea how to distinguish not only as to whether a patient is either drunk or stricken with apoplexy, but also whether he or she may be suffering from the effects of some narcotic poisoning; from what is termed uræmia, a condition in which the blood becomes surcharged with matters which should be excreted by the kidneys; or whether again those conditions which prevail in sunstroke, heat exhaustion, or fainting are present.

If we are able to recognize at once that a patient is suffering from an apoplexy, particularly what is termed a hemorrhagic apoplexy, in which blood from a broken vessel is poured out into the brain-substance, that patient may be so handled as to give him a chance for life and approximate recovery. The treatment is chiefly a negative one, but this negative treatment is of great importance. If such a patient is jarred and moved about, is carried from pillar to post, is twisted and turned in order to remove clothing or for other purposes, the hemorrhage may be unnecessarily increased. In some cases of apoplexy, a comparatively small hemorrhage, from a small blood-vessel, will cause great shock and complete unconsciousness, and the bleeding in such a case may have a tendency to stop if the patient is kept absolutely quiet. Indirectly, too, the danger is augmented by unnecessary movement exciting the heart and general circulation. Even when apoplexy is not of the hemorrhagic variety, quietness should be the rule. On the other hand, if the patient is suffering from drunkenness, or is narcotized by opium or some other powerful drug, it may be all-important to promptly arouse and excite the dormant nervous system; and this is particularly true in cases of narcotic poisoning.

It is by no means easy even for a physician well trained in his art to make a diagnosis in these cases, and therefore a nurse who should fail is not always to be held to rigid accountability. The nurse should, however, be able to distinguish those cases in which the diagnosis is comparatively clear. The rarity of convulsions without heart-failure; flushing of the face or more rarely pallor; stertorous breathing; slow and full pulse;

pupils fixed and one or both dilated; paralysis of one side; retention or involuntary passage of urine; temperature at first lowered, and, as a fatal issue approaches, rapidly rising, are some of the points usually given. These are good, and correct as far as they go, but some details as to the determination of these facts may be given, and a few points may be added. A nurse should know something of the methods of determining whether a patient is paralyzed upon one side. Glancing carefully at the patient's face, usually, but not invariably, one-half of the face will seem to droop, or will be twisted toward the other side; the angle of the mouth on the paralyzed side will perhaps be a little lower, and the lines and contours will be somewhat smoothed out. Lifting the arm, first to one side and then to the other, the paralyzed extremity will fall limp and flail-like, while the extremity of the unaffected side will usually convey a certain resistance to the examiner, even though the patient be totally unconscious. The same conditions can usually be determined by a cautious manipulation of the legs. The respiration is not only heavy and stertorous, but it is frequently also what is called "Cheyne-Stokes" in character,—that is, the breathing becomes less and less marked, descending until the patient does not seem to breathe at all; then, after a few seconds, beginning again and gradually becoming more positive. It is an ascending and descending respiration, —a respiration with a break or gap. Another condition, called conjugate deviation of the eyes and head, very commonly present in cases of one-sided apoplexy, can be determined by simply looking at the patient.

The eyes constantly turn or tend to turn to one side, and the head also shows the same tendency. Usually, the direction is away from the side of the paralysis.

If satisfied that the patient has had an apoplectic attack, or if in doubt about the matter, it is wise, as has been already intimated, to keep him as quiet as possible. The clothes should be loosened so that no undue pressure will be exerted anywhere upon the blood-vessels or internal organs; but great caution should be exercised in the way in which the clothing is handled. It is sometimes far better to cut and destroy clothing than to move the patient unnecessarily in removing or loosening it. Warmth, and even counter-irritation, may be safely applied to the extremities; cold to the head is usually indicated, but sometimes is better not employed, particularly when there is great general depression of temperature. As a rule, a nurse should not take the responsibility of administering medicines in such cases. Certainly emetics should not be given without orders. Probably the only medicine which a nurse would be fully justified in administering on her own responsibility would be a purgative, like one or two drops of croton oil in sweet oil.

Some knowledge of the conditions present and the treatment called for in case of sunstroke or heat-prostration is essential. This affection may be caused not only by exposure to the sun, but by intense heat, particularly with fatigue. The danger is usually very great, and in some instances the temperature rises many degrees. When this is the case every effort should be made to lower the temperature, and this is best done by the use of a cold bath, or by sponging with cold water, or rubbing with ice. An abundance of fresh air should also be supplied. A different plan of treatment has to be employed in different cases. Sometimes the prostration is very great, and stimulants are called for, but these should not be used without medical advice.

It is not always an easy matter to tell that a patient is dead drunk. A patient may smell of alcohol and yet be suffering from an apoplectic attack. He may be drunk and also have a "stroke." In the Philadelphia Hospital a fair percentage of the cases brought in during the apoplectic attack have been drinking, the excessive use of liquor acting as the exciting cause in an individual whose vessels are already diseased. What are some of the indications of genuine intoxication? Consciousness may appear to be lost, but this loss of consciousness is not as absolute as in the majority of cases of apoplexy. The patient can generally be momentarily aroused from his stupor, if it is only to sink back again with a grunt. The face will be more commonly flushed than pale; the pupils will probably be equal, more frequently dilated; the breathing, although heavy, is not strictly speaking stertorous, nor is it likely to be of the Cheyne-Stokes variety. The temperature may be below the normal two or three degrees, but it does not show the peculiar variations which are commonly present in genuine apoplexy. In a case in which, in spite of examination and of studying the points here considered, the diagnosis still remains doubtful, it is better for a nurse to do nothing except quietly take care of the patient and prevent injury. Dr. Reginald

Southey[1] recommends the throwing up the rectum, by injection, of a pint and a half of cold water, with a tablespoonful of common salt dissolved in it. In a case of extreme drunkenness this brought the patient out at once. If carefully administered in a doubtful case, it would certainly be less likely to do harm than stomach-pumping, or galvanizing, a comatose person.

In uræmic coma—that form of unconsciousness from the blood-poisoning which takes place because diseased kidneys cannot properly excrete effete matters from the system—a nurse will not usually be able or expected to determine the condition. In some cases of uræmic coma, however, certain evident physical conditions will enable almost any one to suspect the real state of affairs. Such a patient, for example, may have swelling of the limbs, or of the eyelids or face; a condition of waxlike pallor may be present; the breath may have a urinous or beef-tea odor; the pupils will generally be dilated; and, as a rule, but not without exceptions, one side of the body will not show paralysis more than the other. When a patient is known to be suffering from uræmic poisoning, active purgatives and measures to produce sweating, as warm baths, may do much good.

An attack of fainting is ordinarily not dangerous to life, although sometimes attacks which look like a simple faint are the result of serious disease of the heart or brain. The fainting person suddenly becomes pale, loses consciousness, and usually falls, the pulsations of the heart and the movements of breathing

---

[1] *Lancet*, December 4, 1880.

becoming diminished. In a faint the brain is commonly deprived temporarily of its proper supply of blood. The individual should be carefully placed on the back, and the clothes loosened. In some cases it is well to raise the body and limbs a little above the level of the head. In this way the circulation will be given the best possible chance to right itself. Sprinkling cold water on the face and applying ammonia to the nostrils and heat to the extremities and the stomach will assist in bringing about reaction.

In hospitals in which cases of chronic organic disease of the nervous system are treated, as, for example, in the Nervous Wards of the Philadelphia Hospital, where sometimes two hundred to three hundred patients, a majority of them suffering from structural affections of the brain or spinal cord, are to be found, it is important that the nurse should have clear ideas as to the special care of those suffering from paralysis and other forms of extreme helplessness. Sometimes a paralytic case, and a ward in which paralytics are found in large numbers, are regarded as uninteresting, or possibly unpleasant; but to a faithful nurse, to one who has a just idea of her vocation, these helpless people should be interesting because of their suffering. Certainly, few cases of chronic diseases call for so much true sympathy; certainly, also, in few are the opportunities to comfort and relieve greater. These patients are sometimes literally unable to move hand or foot, or, if the helplessness is not so extreme as this, they are paralyzed either upon one side or in the lower or upper half of the body, more commonly the former. Sometimes sensibility has departed from their limbs. Sometimes they suffer from contractures to such an extent that, as with one case now in the Philadelphia Hospital, they are literally tied into knots. When the condition is not so extreme as this, the limbs may be flexed upon each other and upon their bodies so as to prevent any movement except with great pain. Frequently paralysis of the bowels or bladder, or of the sphincter of the bowels or bladder, is present, causing involuntary evacuation or constant retention of fæcal matter or urine, and adding the dangers of overdistention and even rupture of the bladder.

A quick-witted, efficient nurse will soon learn from observation and contact with helplessness how to be most helpful. In such matters as assisting a case of one-sided paralysis, or of ataxia, in walking, canes, crutches, wheel-chairs, and the like, can be resorted to, and it will be the duty of the nurse to see that they are used in the most advantageous manner. Certain little details of great importance, both with reference to the comfort and cleanliness of the patients and of the rooms, will need attention. When the patients suffer from paralysis of the bowels or bladder, the nurse should bear in mind the attendant evils; should provide against emergencies, and should frequently examine to see that the patients are not in a filthy condition. Many cases of organic nervous disease suffer from inflammation of the bladder, secondary to the paralysis of that organ so often present. It is often necessary that the bladder should be washed out, a work which will usually fall to the physician, and one which a nurse should not undertake without special and explicit directions. She should, however, know how either to wash out the bladder or to assist in doing this if called on by the doctor. Usually a double catheter or some particularly-designed syringe is used in this operation. It may be done, however, by attaching to an ordinary catheter a long rubber tube. The catheter is passed in the usual manner, and then, by means of a funnel attached to the end of the tube, warm water or whatever liquid is directed can be gently poured into the tube, not more than about two fluidounces being poured in at one time. After this has been done, by simply lowering the tube, the water will be discharged from the bladder and run off. This process can be repeated until the discharged liquid is perfectly clear.

Bed-sores are among the most disagreeable and troublesome complications of chronic disease of the nervous system. Unless the greatest care is taken, and even sometimes in spite of close attention and proper precautions, annoying and obstinate sores will form on those parts of the body subject to pressure or attrition, or to the action of irritating discharges. The nurse should, of course, be on the watch to prevent their occurrence; an ounce of prevention in such cases being emphatically worth a pound of cure. Among the most useful means of prophylaxis and cure, I need only mention the use of the water-bed and of air- and water-cushions.

Small pillows may sometimes be used. When rubber air-cushions are employed they should be smoothly covered, and pins should not be used in the covers; preferably they should be sewed together. When a

c

water-bed is used, it is well to put an old blanket or a cloth of some kind under it to keep it from sticking. The water which is used to fill it should be at a temperature of 70° F., and should be renewed every ten days or two weeks, or oftener.

Charcot, the distinguished French physician, has pointed out the fact that occasionally, in cases of cerebral hemorrhage, a few days after the "stroke" an acute bed-sore or eschar will form on the buttock of the paralyzed side, long-continued pressure having nothing to do with its production. In rare instances the sore occurs on the sound side. In a case of this kind, usually at first, an erysipelatous blush makes its appearance, in the centre of which a vesicle is soon seen, and this breaking down, a formidable sore rapidly results.

Plans of treatment for bed-sores without number have been tried and recommended. Brown-Séquard strongly advocates the use of alternate hot and cold applications. For ten or fifteen minutes daily, sponges soaked in hot and cold water can be thus employed, or ice bladders alternating with hot poultices may be applied. The effect is to stimulate the circulation, and to promote the formation of granulations. The simple water dressing, the use of carbolized oil, and washes of carbolic acid and potassium permanganate, stimulation with solution of the nitrate of silver and sulphate of copper, packing with iodoform powder, the application of charcoal poultices and of various ointments, and powdering with oxide of zinc, and other methods used singly, or some of them in combination, are doubtless all more or less familiar to nurses who have had any experience.

A good plan to prevent the development of bed-sores is to use alcohol, or the fluid extract of hamamelis, as a wash. It is also important to watch and turn the patient from side to side.

Some years ago I experimented in the Philadelphia Hospital with a method of treatment first suggested by Spencer Wells, of England. This I consider very valuable. It consists in taking a piece of fine silver plate and cutting it to the size of the bed-sore. A piece of zinc of the same size is taken, and the two are united by a wire. The zinc plate is laid a little above the sore, and the silver plate is laid on it, and the two are secured by adhesive strips. I have known sores, treated in this way, fill up with granulations in two or three days.

A good nurse can do much to assist the physician in the treatment of sleeplessness from any cause. A fussy, a talkative, or a thoughtless nurse will be likely to make such a patient worse. The manner in which the evening is spent will often have much to do with the production of sleeplessness or the induction of sleep. Even when hypnotics are administered for the sake of causing sleep, the way in which this is done, and the course which is pursued subsequently, will have a decided bearing upon the results. Of what avail will be the drowsy syrup, or the soothing draught, if, after its administration, the nurse becomes noisy and excites the patient?

In delirium tremens, sleeplessness and suicidal impulses are common. The pulse may be very feeble and the nervous prostration and inability to take food extreme. Nourishing food must sometimes be given by force, and such patients must always be closely watched.

In meningitis, in tetanus, in delirium tremens, as well as in fevers, and in some forms of insanity to be hereafter considered, the treatment of delirium will call for the intelligent help of the nurse. No matter to what the delirium is due, quiet and peace are always essential. Usually the room should be kept not only quiet but to some extent darkened. Care should be taken not to unnecessarily jar the bed. Loud talking, laughing, or any form of noise, should be interdicted. In brief, nothing should be done to startle or excite. In acute alcoholic delirium the mental condition of the patient is sometimes one which requires extreme care. Terrible anxiety with frightful hallucinations of sight and hearing are present. Sleeplessness is continuous. The patient must be vigilantly watched, must be soothed and comforted, and if necessary must be restrained. When delirium becomes so violent that it is absolutely necessary to use temporary physical restraint, the greatest possible care should be exercised. Physical restraint is indeed seldom necessary, and should always be avoided if possible. A patient can sometimes be wrapped in a sheet or blanket for a time.

It is not within the scope of the present chapter to go into details as to food and its preparation. Such details belong more properly to a work on dietetics. I will only say that in the management of nervous patients of different classes attention to diet is often of the greatest importance. Sometimes, as in the so-called "rest treatment," a special milk diet is ordered, or a diet in which the patient is restricted to certain articles prepared in certain ways. The orders given should be carried out explicitly. If the milk is to be peptonized, it should be exactly in the manner directed. If skim milk is ordered, skim milk should be given; and if raw-meat soup is directed after having been prepared in a certain manner, it should be prepared in this way and in no other. I wish to impress the importance of fulfilling every detail of treatment with care and exactness.

The management of patients addicted to the opium or other narcotic habit requires peculiar qualities in a nurse. Firmness is here of the utmost importance, and vigilance cannot be too great. The victims of these unfortunate habits are weak in resolution beyond all other sick people. It may be laid down as a general rule that they can never be fully trusted. The majority of patients, says Kane, " express their willingness to be rid of their habit, and do endeavor up to a certain point to assist themselves, but at this period will-power, naturally weakened, gives way, and good resolves are thrown to the wind. It is at this time that every facility for full control of the patient is necessary, for without it the sufferer will invariably stop treatment, claiming that the suffering is beyond his strength, bemoan his sad fate, and return to the old habit with renewed force, exclaiming with Coleridge's son,—

'O woful impotence of weak resolve.'"

Such patients must be ceaselessly watched. They must not be allowed to come in contact with servants and others through whom they can buy the drug. Often they may have the narcotic concealed. The nurse is in these cases the physician's main-stay; if faithless, treatment will invariably fail.

In the course of the treatment of narcotic habits certain periods may arise in which it is important to exercise the utmost care in the management of the patient, and a nurse should be well informed with reference to this matter. In the method of treatment advocated by the German physician, Levenstein, for example,—that of rapidly reducing the amount of the narcotic taken,—a condition of collapse or semi-collapse is likely to come on about the time the drug is entirely removed, and when this occurs it is necessary that the patient should be closely watched and attended to by the nurse as well as the physician.

CHAPTER II.[1]

Massage—Movements—Muscle Beaters—Bathing—The Revulsor—Surface Thermometers.

A FEW years ago it was difficult, even in large cities, to get men or women who were capable of properly administering massage to patients; indeed, the number of masseurs and masseuses, skilful and unskilful, was for a long time quite limited. The times have changed in this respect. Every large city now contains many individuals who are, or who claim to be, capable of giving this treatment. Often it is simply a claim. Not a few imagine that it is only necessary to have a little strength and more assurance to be able to treat a patient by the method of massage. In consequence much evil results; not only to the patients, but to the physician who recommends the treatment, and to the cause of massage itself. It is not uncommon for nurses who have managed perhaps two or three patients, and while caring for them have watched a skilful masseur or masseuse operate, to imagine themselves capable, without special training, of undertaking massage in all its forms.

No one should set up as a practitioner of massage without having received special instruction. I have

---

[1] In the preparation of this chapter I have received some suggestions from Mr. William A. Kerkhoff, a well-trained and skilful Philadelphia masseur.

known an individual to undertake to administer massage for pay after observing the operation on one or two occasions. Instruction should be received, and care should be taken that it is of the proper kind. A nurse to-night cannot be a masseur or masseuse to-morrow simply by thinking over the matter. Both teaching and preliminary practice are requisite to make a good operator. The amount of time required to thoroughly learn the art depends upon the intelligence, tact, and manual skill of the individual as well as upon the kind of teaching and the amount of practice.

It is not absolutely necessary that masseurs should have a comprehensive knowledge of anatomy and physiology, but they should know something of these branches of medical science; they should know the names, positions, and modes of action of the principal muscles and groups of muscles of the body; they should know something of the size and shape or contour of these muscles in order that they may be better able to isolate and treat them; they should know something of the skin and its functions and powers.

In my own practice, both hospital and private, I have made it a point to study the methods of different operators, and I find that these differ very much. In Philadelphia, the majority of those who attempt massage have methods which are, in some respects, wrong, although we have here a respectable minority of skilful operators. I have not only watched the procedure myself, whenever opportunity offered, but have also questioned my patients as to their own feelings and observations. To a majority of nervous patients, massage is pleasant; some care but little for it; to a few it seems to be absolutely intolerable. When, however, a patient complains that she does not like or cannot endure the procedure, I always make it a point to observe the method pursued. I have had patients who were rendered more nervous by massage administered by one person to speak of it as almost a luxury when they have received the treatment from the hands of another. One of my patients, an intelligent lady, had in succession, although not by my orders, five different masseuses. Her rather amusing criticisms upon these individuals serve to point out some of the things to be avoided. She said that the first teased and tantalized her; the second wrung and twisted her; the third mauled and bruised her; the fourth was too dainty; and the fifth almost talked her to death. The patient may have been a little hard to please, but in the main her criticisms were just.

Murrell[1] says that the word massage is derived, according to some authorities, from a Greek word meaning to "rub," while others refer it to the Arabic word *mass*, to press softly. Dr. Benjamin Lee[2] says that it comes to us from the Greek through the French, and means simply "kneading," the idea to be conveyed being that the operator works the flesh as the baker works his dough. He holds correctly that it is better to use the French word than to attempt to translate it into

---

[1] "Massage as a Mode of Treatment."
[2] "Tracts on Massage." Translated from the German of Reibmayr, with notes, Philadelphia, 1885.

English, because the English equivalent is used to describe one of the particular modes of massage, and also because the former has now a well-established position and definite signification in scientific medical literature all over the world. The same authority defines massage as a communication of motion to the tissues of the living human body from an external source for therapeutic purposes; but this definition does not define the term as it is commonly used,—that is, under it might be included Swedish movements and motion communicated by means of steam, electricity, or special mechanical agency. The latter forms of massage are sometimes spoken of as *mediate*,—that is, they are motions or movements communicated to the body through some contrivance or machine. It is to what is sometimes called *immediate* massage that the term is commonly restricted, and it is with reference to this form that my remarks in this connection apply. Massage, then, in the sense in which I here employ the term, is the manipulation by a definite process of the tissues and organs of the human body by a living operator, for the purpose of promoting health or of relieving or curing disease.

The subject has been unnecessarily confused and complicated by the multiplications of terms describing so-called methods of massage; thus, Estraderé, according to Lee, speaks of such a variety of procedures as stroking, friction, kneading, sawing, pulling, pinching, malaxation, percussion, chopping, clapping, pointed vibrations, and deep vibrations. These terms really represent only a few essential methods of procedure.

The best authorities, as Mezger, Berghman, and Helleday, Lee, Murrell, Jacoby, and others, describe four essential methods of procedure, namely, (1) stroking, or *effleurage;* (2) friction, *massage à friction;* (3) kneading, or *pétrissage;* and (4) percussion, or *tapotement.* Understanding these four methods practically, it will be found that the other so-called varieties are simply modifications of these, or are not strictly forms of massage, but come more properly under the head of exercise or movements.

My special purpose in this chapter is to give a few hints and suggestions as to proper and improper methods of performing massage. For details and descriptions of the procedure, the writings of Lee, Murrell, Jacoby, and others can be consulted. I will, however, for the sake of a clear understanding of what is said, quote the essential portions of the descriptions by Jacoby[1] of the principal kinds of massage.

"Stroking is nothing more than a very light centripetally-propelled pressure. It is executed in the following manner: The volar surface of the ends of the fingers, or of the entire hand, having been applied to the part to be massaged, at a point situated more peripherally than the affected portion, is then pushed centripetally forward and a short distance beyond that part. When this hand has reached its destination, the other hand is placed at the starting-point and the same movement executed. Meanwhile the first hand has

---

[1] "Journal of Nervous and Mental Diseases," vol. xii. No. 4, October, 1885.

been brought back, so that by the time the second one has fulfilled its purpose it is ready to begin again. This is then repeated with regularity. The time to be devoted to each stroke will vary very much, the strokes also following with more or less rapidity. The amount of pressure to be applied is not to exceed that of the dead weight of the hand. Effleurage of the extremities may also be executed in such a manner that the thumb and first finger are widely separated like a V, and the extremity grasped between them. Mosengeil advocates the use of a certain amount of pressure in this manœuvre; in fact, he recommends varying the pressure from moment to moment, so that it becomes undulating in form."

"*Massage à friction*, or rubbing, is a forcible pressure with a concomitant motion of the hand forward. This manipulation may be executed in various manners. The points of the fingers having been applied to the affected part, pressure is exercised, and they are moved in large or small circles, or simply to and fro, the pressure being maintained. For this purpose the ends of one or more fingers may be employed. If a particularly great amount of force is desired, the first phalanx of the thumb of the left hand may be applied to the spot, and pressure exercised upon its dorsal surface with the thumb of the other hand, the rubbing then being executed with both together. By this means we are enabled to act upon deeper-lying structure. Rubbing and stroking ought to be combined in such a manner that the operator simultaneously executes the one with his right hand and the other with his left."

"*Pétrissage*, a kneading in the true sense of the word, may be applied wheresoever soft, graspable tissues are present, and very intense effects may be attained by it. The part to be acted upon, generally a muscle, is grasped between the thumb and index finger, or between the thumb on the one side and the four fingers upon the other. It is then isolated as much as possible and lifted out of its bed, at the same time pressure being exerted upon a certain part. Then those parts more centrally located are taken, until the entire muscle has been acted upon. For this purpose either the points of the fingers or the pulps of the terminal phalanges may be used. If the points of the fingers are employed, the procedure is certainly rendered more energetic, but at the same time more fatiguing for the operator.

"Pétrissage may also be rendered more energetic, when acting upon large parts, by using both hands."

"*Tapotement* is simply percussion of the affected part, either with the palm of the hand, the fist, the curved finger, a percussion hammer, or other instrument specially adapted for the purpose. Percussion without instruments is, as a rule, performed with the points of the fingers, these being semiflexed, and the movements of the percussing hand taking place at the wrist. It is this form which is most applicable over superficial nerves. When it is deemed desirable to act more energetically upon deeper-lying structures, generally large muscular groups, the wrist-joint must be stiffened, the fingers held firmly extended, and motion allowed to take place at the elbow and partly at the shoulder-joint, the hand of the operator thus meeting the body of the patient at right angles."

The terms which are used to designate those who follow massage as a business, and also the methods and processes of massage, are somewhat awkward for English-speaking people, more particularly for nurses and operators who have not, as they cannot be expected to have, a French as well as an English education. The proper term for a male operator is *masseur*, for a female, *masseuse*, the plurals being *masseurs* and *masseuses*. The verb which expresses the performance of the procedure is *masser*. It is therefore proper to speak of a patient as being masséed, or of massécing a patient. I would not advise the use either of massageing or of "massacreing," although the former is sometimes used by good authority. The latter is certainly expressive in describing the violent performances of some of the untrained exponents of the art. It is common to speak of the process as "rubbing," "kneading," or "manipulating," and perhaps the use of one or the other of these expressions cannot always be avoided; but as they convey double meanings, or do not really express in English what is wished to be said, it is better to use the French terms until some more exact English equivalent is suggested.

As in making electrical applications, so in using massage, attention should be paid to the position of the muscles and of the limbs. The muscles should not be tense; the limbs and trunk should be so placed, or so supported, as to be in a position of relaxation: usually this can be accomplished by having the limbs, as recommended by Graham, midway between flexion and extension. A muscle which is tense and contracted cannot be thoroughly manipulated.

A certain knack or tact is soon acquired by a good masseur or masseuse. In most cases wrist and hand movement should be employed. Just as it is important at the start for the young physician to obtain what is called the wrist movement in percussing patients, so it is essential that any one attempting massage should understand these wrist movements. Patients should not be mauled and twisted by movements which call into play the muscles of the arms and trunk, when the whole procedure can be better accomplished by the forearms and hands.

Wherever it is possible, the operator should learn to use both hands at once. If the left hand, as in so many cases, is comparatively useless, efforts should be made to train it especially, using it in preference to the other hand until the two become equally skilful. To be ambidextrous or either-handed, is a great advantage in a masseur.

Should the massage be dry, or should grease or liniments of some kind be used? Differences of opinion about this practical matter prevail among those who have taught and written upon the subject. Mitchell, for example, advises, or at least permits, in some cases, the use of cocoa-oil or vaseline, but says that these make the work less efficient and more difficult. He does not order them unless it is considered advisable to rub into the system some oleaginous material. Busch and some other writers strongly recommend the employment of oil, liniments, or ointments; Murrell as strongly favors the dry method, saying that the less ointment that one uses the better, and that vaseline is never admissible. The only, or almost the only, exception he makes is when the patient suffers from some form of specific disease, when the operator should use an antiseptic preparation, carbolic acid or oil of cloves and lard, for example, for his own safety and protection. My own view is that, as a rule, nothing in the form of oil, grease, or liniment should be employed. A skilful operator will usually do better without than with anything of this kind.

Some attention must be paid to the direction in which massage is carried out,—that is, whether from the periphery towards the centre of the body, or from the centre towards the periphery. As a rule, the operation should be " from the extremities to the trunk, from the insertion to the origin of the muscles, and in the direction of the returning currents of the circulation" (Graham). The rule, however, is not invariable, but the masseur is not to be supposed to have sufficient knowledge of the subject to decide. The physician should decide; but when no special orders are given, it is proper to begin at the extremities—at the fingers or toes—and proceed up the limbs, finishing with the muscles of the back, chest, and abdomen.

A peculiar twisting or wringing movement used by some operators should be especially avoided. In manipulating the fingers, for instance, instead of supporting the under surface of the finger and masséeing by causing the ball of the thumb to adhere to the muscle by downward pressure and then gently rotating, a twisting spiral movement is performed around the entire finger. In the same way the thumb or the hand is made to perform twisting or pulling or separating movements across the larger muscles, instead of pinching, kneading, or rolling them skilfully. Sometimes the process is simply a form of irregular friction of the skin instead of being true massage.

It is of vital importance that those who follow massage as a business should keep in good physical condition. If the number of patients treated is small, a little extra exercise may be taken. On the other hand, too many patients should not be undertaken in a single day. Probably six or seven patients are as many as can be done justice to by an operator in good physical condition; but, of course, the exact number will depend in great measure upon individual skill and endurance. If too many patients are treated, those who come last will be likely to be slighted, and nothing scarcely will prove more injurious to the business of the operator than this. My experience is that patients pay close attention to the directions of the physician as to the length of time which is to be devoted each day to the procedure. In this, as in so many other matters, patients wish to get the worth of their money. They soon learn whether some part of the body is not receiving proper attention.

It requires considerable strength and vigor to be a good masseur or masseuse; but better even than strength and vigor is the ability to use what strength the individual possesses to the best advantage; in other words, skill. The mistake is made by some operators of supposing that vigorous or even violent treatment is required. I have heard the practitioners of massage of a certain city, which I will not name, spoken of as greasers and bruisers, and while I do not think this is deserved by them as a class, the practice of a few gave some reason for the derision. Patients should not be roughly handled. In some cases after the first few treatments, if the individual has had a delicate skin or some peculiar state of the circulation, some marks and bruises may be left upon the body, even when the operation has been performed skilfully and not improperly; but patients should not appear after such an operation covered with bruises and blotches as if they had fallen down a precipice or had been to Donnybrook Fair. I have known patients comparatively strong complain bitterly of the severity of the treatment to which they have been subjected.

On the other hand, it is possible for the treatment which is given to be too mild, or to be slighting in character. A patient should always be thoroughly treated, should be gone over systematically. One part of the body should not be treated with vigor and the other slighted. An operator who is not very strong, or who attempts to do too much work in one day, will sometimes start out vigorously enough, but before getting through will become fatigued, and in consequence the part of the body which receives attention last will also receive it least. The operator should so gauge time and strength as to be able to carry out the entire treatment uniformly. Treatment should, if anything, end with more vigorous movements than are employed at the start.

I may be pardoned also for making a few remarks as to the mental requirements of masseurs. They should be discreet and judicious, and should have that personal quality which is best expressed by the term "common sense." Inquisitiveness is to be avoided. They, like the doctor, may be called upon sometimes to penetrate the privacy of families, and they should first, before all, learn to mind their own business. Where more than one patient is in a house or hospital, it is well not to pry too much into matters relating to other patients than the one being treated. Talking too much should be avoided. Some practitioners of massage are as garrulous as the traditional barber, and are as much of a nuisance to their victims. It is not necessary, nor to be advised, that they should be taciturn or morose in the presence of patients. A little pleasant talk will do no harm, but they should avoid undue talkativeness; they should particularly avoid gossiping and boasting of their own powers. They should not set themselves up as "magnetic healers" or "professors," being particularly careful not to impress either doctor or patient as to the amount of electricity they can convey by their manipulations.

The masseur should particularly refrain from talking much in the evening, so as not to charm away sleep. The lights should be turned away from the patient's eyes when the seance is in the evening.

The patient should be in a well-ventilated room, say of 70°. The chest should be elevated somewhat, so as to promote free and easy respiration. Each limb should be covered after being treated, and the masseur should not unnecessarily expose any part of the patient. The masseur or masseuse should refrain from harassing the patient's mind by talking of his disease. The physician should in some cases inform the operator as to the purpose for which the massage is applied, as this will give the masseur more confidence and make more certainty as to the thorough performance of his work.

The treatment should be commenced very gently, gradually increasing in strength as the patient becomes more able to bear it. Care should be taken not to overdo the treatment on the first attempt. The masseur should exercise his judgment as to the amount of vitality and physical strength possessed by the patient, and upon this in part the treatment should be based.

Movements, in the medical sense, are certain motions and operations performed for the purpose of helping the diseased human body. They are, as Taylor expresses it, motions of specific kinds, having specific effects, practised for specific purposes, and intended to secure definite results. Cumbersome and complicated apparatus is not required to carry out the movement treatment. Much can be done with no machinery, although appliances, which have a certain value, have been invented. My remarks will have reference chiefly to procedures which do not call for any special apparatus.

By writers and teachers various divisions and subdivisions of movements have been laid down. Efforts have been made by Ling and others to designate and classify both positions and movements; in other words, to establish a terminology for the movement cure. It is not my purpose, however, to go into this branch of the subject at present.

Movements are sometimes spoken of as active and passive, or as single and duplicated. Active movements are those more or less under the control of the individual making or taking part in them, and they are performed by him under the advice or direction, and sometimes with the assistance, of another. They proceed from within; they are willed. Passive movements come from without; they are performed on the patient, and independently of his will. He is subjected to pushings and pullings, to flexions and extensions, to swingings and rotations, which he can neither help nor hinder. The same movement may be active or passive according to circumstances. A man's biceps may be exercised through his will, against his will, or without reference to his will. A single movement is one in which only a single individual is engaged; speaking medically, single movements are those not executed by the physician or attendant; they are, of course, active. Duplicated movements require more than one for their performance. Duplicated active movements are those to which I most commonly resort in the treatment of nervous disorders. In these the element of resistance plays an important part. The operator, with carefully-considered exertion, performs a movement which the patient is enjoined to resist; or the latter undertakes a certain motion or series of motions which the former, with measured force, resists. Skill, tact, and experience are here of great value, in order that both direct effort and resistance should be carefully regulated and properly modified to suit all the requirements of the case. By changing the position of the patient, or the manner of operating on him, from time to time, any muscle or groups of muscles may be brought into play. It is wonderful with what ease even some of the smallest muscles can be exercised by an expert manipulator.

To impress what is meant by duplicated active movements, I will illustrate them by supposing two cases: one, a case of partial paralysis of the muscles of one arm, the other a patient with hysterical paraplegia. In these cases muscular exertion can, to some extent, be called forth, the paralysis not being absolute. Suppose it is desired to bring into activity the biceps and brachialis anticus, the muscles particularly engaged in flexing the forearm. The arm being extended, with the palm of the hand upward, the manipulator takes hold about the wrist and directs the patient to draw the hand toward the shoulder. As the latter performs this movement, the former carefully resists, gradually, however, allowing the hand to reach the shoulder, where the patient tries to keep it, while the operator now slowly brings the arm again to its extended position.

To act upon the muscles of the trunk, the patient is seated upon a stool of the proper height, without back or arms, and is directed or caused to bend forward or backward or sidewise, or to semi-rotate the body, or to perform any other available movement, the operator antagonizing the patient and the patient the operator. The arms may or may not be made to take part in these movements, being, in the former case, elevated, flexed, extended, etc., in unison with the movements of the trunk. If it is wished to exercise any or all of the muscles of the lower limbs, the process is obvious. Beginning with the toes, these being flexed, while the patient endeavors to extend them the operator resists his efforts, or the procedure may be reversed, the patient offering the resistance. Similar methods are adopted for calling into play and developing the extensors, abductors, adductors, etc., of both feet and legs.

These duplicated active movements are designed to call out the latent power in the muscles of the individual. They are called duplicated active movements because the operator and the patient take part in them, and the important practical point is to see that both do take part. Great skill is acquired by some operators in this matter. The various movements of the trunk and limbs should be carefully studied, and the masseur should learn by practice the neatest and least awkward method of opposing the patient's movements or causing the patient to resist the force applied. As in massage, observation of two manipulators will reveal the greatest difference in this respect: one will be quick, easy, graceful, and encouraging; the other will be slow, stiff, awkward, and disheartening. The force which is used should be carefully graduated to the strength and particular conditions of the patient. The movement which it is desired to bring out and improve should not be too forcibly resisted. Patients can be helped on to a cure much more rapidly by carefully coaxing out their latent energies.

The muscle-beaters of Ruebsam may sometimes be directed to be used by the physician. They are appliances for beating any portion of the body. In using them particular attention should be paid to the explicit directions of the physician as to the amount of force to be used, and the portion of the body to be treated.

Fig. 1.

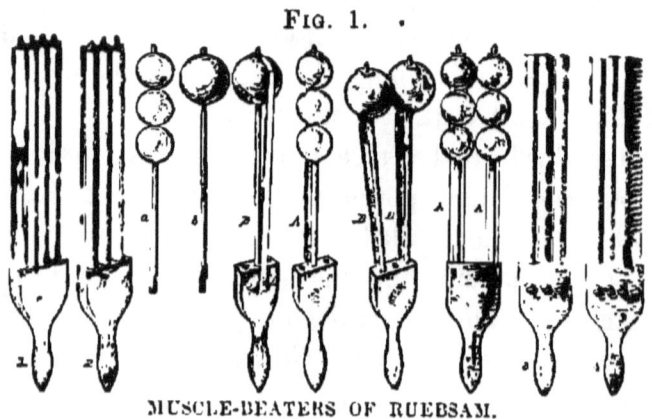

MUSCLE-BEATERS OF RUEBSAM.

The muscle-beaters are shown in the above cut,—No. 1 with four, and No. 2 with three fingers, and *a* and *b* as attachments. The tubing may be removed from Nos. 1 or 2, and the sticks with the balls fixed in their place, single or double, as shown in figures *A* and *B*, and *AA* and *BB*, to answer for different parts of the body.

The bones in Nos. 1 and 2 are made of rattan, which fit loosely in the sockets so as to move forward and backward, to make the stroke flexible. The turning could easily be prevented by running the screws into the sticks, but that would not only keep the sticks from moving, but they would also break or bend very easily, which the turning prevents.

The rubber tubing, and also the balls, in place of the flesh or muscle, make a good cushion or beater to overcome any bruise, even from a heavy blow. Elastic straps hold the entire hand together and allow the fingers to bend and move over the most delicate parts of the body without fear of injury. The beaters Nos. 3 and 4 are still more flexible, as the bones are represented by spiral springs, over which slide rubber tubes; but they cannot be changed, as the springs are fastened in the handle.

In the care and treatment of nervous patients bathing, either general or local, is frequently employed. In text-books on general nursing and special works on hydro-therapeutics details of the different forms of baths and special procedures will be found. I may, however, say a few words about certain special forms of bathing. In the treatment by rest, seclusion, massage, and electricity it is nearly always advisable that a sponge-bath should be given early every morning. It should be done quickly and carefully. The sponging should be downward; the addition of a little alcohol to the water will sometimes make it pleasanter and more desirable where it is intended to get a cooling effect. The sponging should be cold or tepid, according to the directions of the physician; usually tepid sponging is to be preferred, particularly in the early period of treatment. It is a well-known rule founded on physiological principles that baths should not be taken immediately after meals. The sponge-bath, however, may be taken after a cup of milk, cocoa, or coffee in the morning, or it may be given as other baths are administered between meals or at bedtime. Even the sponge-bath should not be given when the stomach is full of food.

Dr. Mitchell says that for some reason the act of bathing, or even being bathed, is mysteriously fatiguing to certain invalids, and if so the general sponging should be done for a time but thrice a week. With reference to the matter of the temperature of the water to be used, either sponge or other bathing, the nurse should be observant, and should report the effects to the physician so that he may be guided in his direction.

In the treatment of some forms of delirium the wet pack is of great value, but should be used with care.

In the treatment of some spinal disorders, as myelitis, and also in the treatment of neuritis, spinal irritation, and some other nervous affections, the alternate application of very hot and very cold water, or of hot water and of ice, is resorted to and must be given into the hands of the nurse. This treatment is carried out in one or two different ways. One method is to have a basin full of hot water, and another with ice-water and ice broken up in it. A sponge is dipped into the very hot water, is squeezed out quickly, and rapidly passed up and down the spinal column or limbs; then another sponge is dipped into the cold water and the same procedure is used. In this way applications first of the hot water and then of the cold are made, usually from ten to twenty minutes, but according to special directions. Care should be taken in using these local baths not to unnecessarily expose, and not to have the water dripping or running all over the patient. As soon as the operation is completed the patient should be rapidly dried and be made comfortable. If the operation causes an undue amount of shock, that fact should be reported at once to the physician for further advice.

In an article by Allan McLane Hamilton, M.D., on "The Use of Revulsives in Diseases of the Nervous System," published in the *Philadelphia Medical Times* for September 4, 1875, is described an ingenious instrument for the alternate application of dry heat and dry cold, a most valuable form of revulsion in spinal irritation and other nervous maladies. It consists of two chambers of brass, three inches in diameter by one and a half inches deep. These have screw plugs inserted, so that they may be removed and the chambers filled, one with cold salt and water, the other with hot water. These chambers are fixed on a rod and separated by an insulating or non-conducting substance. The rod terminates in a handle. The flat surface, covered by thin flannel, is placed against the bare back, on either side of the spinous processes of the vertebræ, and the instrument moved up and down quite rapidly. As the heated surface moves instantaneously to where the cold one was the instant before, the effect is quite marked.

Physicians who have charge of cases of nervous disease and of insanity, to some extent resort to the use of surface thermometers, chiefly for purposes of diagnosis; and it is important that nurses should have some special knowledge with reference to the construction and methods of handling these instruments. They are not used in exactly the same way as the thermometers which are employed for taking temperatures in the mouth, axilla, or rectum. The surface thermometer is an instrument for determining local temperature, which is often a matter of much interest, and not rarely of practical importance, in the diagnosis and treatment of nervous diseases. The surface thermometer can, for instance, be used with advantage in infantile paralysis, in some local palsies, in hemiplegias, and paraplegias. In certain cerebral disorders it will sometimes show a higher temperature on one side of the head than on the other. In local spasmodic affections and neuralgias it may also indicate delicate thermal changes. It may likewise be brought into service for other diseases than those of the nervous system; and in surgery, ophthalmology, dermatology, etc., it has its uses; but I am concerned at present only with its employment in neurology. We owe it chiefly to the elder Seguin, who, a few years ago, urged on the profession the necessity of an instrument of this kind. The bulb of the Seguin thermometer is flat so that it will rest steadily upon any part of the body. It is carefully graduated and very sensitive.

As we have no "norme" or normal temperature for surfaces, we make our observations by comparison of diseased with healthy parts. If we have only one thermometer, we place that instrument first over the affected muscle or organ, and then over an analogous or corresponding healthy region, noting the differences of temperature. Generally two or more surface thermometers are employed at the same time. Having two exactly alike, when it is desired to make observations, they are first warmed to three or four degrees below the healthy standard, and then applied perpendicularly and without marked pressure to the skin. The places of application will depend upon the nature of the disorder. If dealing with a local paralysis of the face, arm, or leg, one instrument can be applied to the surface over the muscles affected, while the other is held to a corresponding point on the healthy side. The two temperatures are noted and compared.

In the Seguin thermometer the base of the bulb is comparatively broad and flat, and the thin glass surface forms an elastic diaphragm, which, by yielding to pressure, introduces an error. In two thermometers examined by Dr. H. Leffmann and myself the error from this cause amounted to 0.5 and 1, respectively. Dr. L. C. Gray, in his experiments, used Seguin thermometers constructed especially to avoid this error.[1] In a discussion at the Philadelphia Pathological Society,[2] I explained a modification of the Seguin thermometer devised by Dr. Leffmann and myself. Two corks were placed in a common test-tube, one in the middle and the other at a level with the mouth; and through these the stem of the thermometer was passed, the bulb being brought flush with the outer surface of the lower cork. In this way currents of air, and the effects of handling and of pressure, were in great part avoided.

The Mattson's Surface Thermometers, obtained from

---

[1] "Cerebral Thermometry," *N. Y. Med. Jour.*, August, 1878.
[2] *N. Y. Med. Rec.*, December 14, 1878.

J. H. Gemrig & Sons, Philadelphia, are reliable in most respects. The bulb consists of a spiral tube, about 1 millimetre (one-twenty-fifth of an inch) in diameter and 5 centimetres (two inches) long, making nearly three turns. The stems in different instruments vary in length from 13 to 15 centimetres (five to six inches). The mercurial column is very fine, probably about 0.33 millimetres (one-seventy-fifth of an inch). The scale runs from 90° to 110°. The bulb and lower fifth of the stem are protected by a cylindrical casing of hard rubber, open below. The stem passes through a rubber diaphragm within the casing. The upper part of the case unscrews. The instruments are not appreciably influenced by pressure. This is due to the spiral bulb, which presents a firm cylindrical, and also extended surface. The encasing carries out the above modification by tubes and corks. It is necessary to enclose more than the bulb. We would suggest a cover of felt, closely fitting the bulb and covering only a small portion of the stem. Experiments made with covered and uncovered bulbs have shown a difference of nearly a degree in the record.

CHAPTER III.

Forms of Electricity—Toepler-Holtz Electrical Machine—Faradic Apparatus—Galvanic Apparatus—The Mechanism, Management, and Care of Faradic and Galvanic Apparatus—Hints, Cautions, and Contraindications in Using Electricity—The Milliampèremetre—Electrodes—Conducting Cords—Methods of Applying Electricity.

IN the present chapter I shall call attention to the use of electricity by nurses; but while I believe nurses should sometimes be allowed to use this agent, I wish to have my position with reference to this matter clearly understood. As a rule, only physicians, or others under the special direction of physicians, should use electricity for medical purposes. Nurses, therefore, should be permitted to make use of it only under special orders for particular objects. In order to employ electricity in all its forms, and for all or even many of its legitimate purposes, no one is qualified thoroughly unless he or she has received a regular medical training. Certain dangers, to which I will recur later, attend the use of electricity in the treatment of disease; and some of these dangers are of such a character that to avoid them presupposes a knowledge of the physiology of the nervous system and of the special senses. In brief, my view is that in a certain limited number of cases, under the direction of the doctor, the well-trained nurse may be entirely competent to use electricity. This is the view that is held by some of our best neurologists, who allow the nurses who have charge of certain of their patients to make electrical applications which do not involve any danger or require any unusual knowledge. It is well, therefore, that the nurse should know both what she can do and what she ought not to attempt. It is a serious mistake for any one not properly educated in medicine to set up as an electrician. I have known of one or two instances in which, after receiving a few directions from a physician, and after getting a few hints from a mechanical electrician, the nurse has imagined that she had the whole science of electro-therapeutics at command. Woe to the unfortunate patient who falls into the hands of such an electrician!

It is not necessary for the nurse to know much about the physics of electricity,—that is, about the nature of this wonderful force or about the principles involved in the construction and action of batteries; neither is it requisite that she should have a profound acquaintanceship with the physiology of electricity,—that is, with the science which treats at length of the action of this force upon the animal tissues and organs. She should, however, at least understand that she is dealing with an agent which is capable of doing harm as well as good, and which, therefore, requires great care in its management.

In the practice of medicine to-day, three forms of electricity are chiefly used. These are termed faradism, galvanism, and franklinism. I insist on these terms. It, perhaps, makes little difference scientifically whether a certain form of electricity be designated as faradism, or galvanism, or franklinism, or something else; but by always adhering to these names, and by always applying them in the same way, any one can understand the recent books on electro-therapeutics, and also the language and directions of most physicians. These terms are derived from the names of men and involve no theories, and they are explicit enough for practical purposes. The faradic machine is often improperly called a galvanic battery; it is sometimes properly spoken of as an "induction coil," or an "electro-magnetic" apparatus. The galvanic battery is often called the "constant" or "continuous current" battery, or the "voltaic" battery, and so on through a long list of terms which are not in themselves wrong; but it is better for the sake of simplicity to drop all terms but faradism, galvanism, and franklinism, and their derivatives, faradic, galvanic, etc.

I shall dismiss franklinism with a word or two. This is the form of electricity which is derived from the static or frictional machine. What is known as the Toepler-Holtz electrical machine is the one commonly employed. It is used by specialists to a limited extent in office practice. A nurse or attendant may be required occasionally to assist the physician in making applications of franklinism,—to arrange the insulated stool, to revolve the plate, to assist in adjusting the condensers, or even in part, perhaps, to make the applications. It is best not to use electricity of this kind except in the presence of a physician or under his special directions. An improved Toepler-Holtz electrical machine is shown in Fig. 2.

With faradism the nurse will probably have more to

Fig. 2.

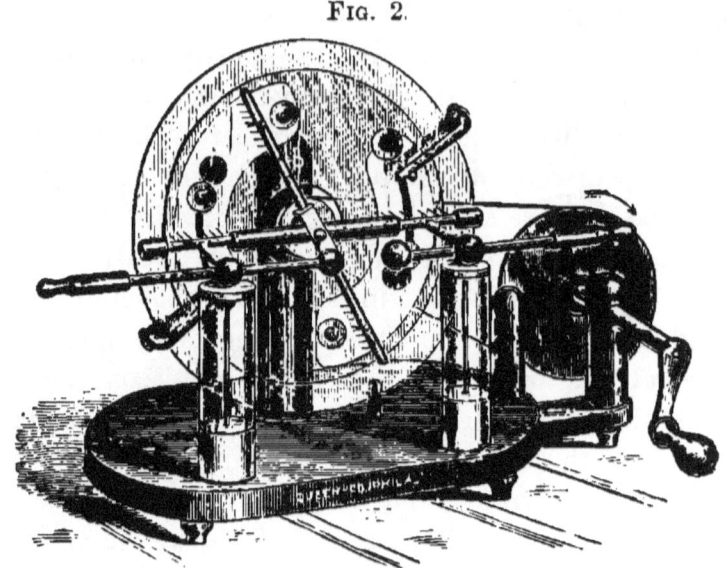

TOEPLER-HOLTZ ELECTRICAL MACHINE.

do than with any other form of electricity. A faradic battery, machine, or apparatus (Fig. 3) usually consists of a cell and a double coil. Some faradic machines are made with several coils, but most of them have but two, a primary and a secondary. These coils are unfortunately not shown in this instrument; but one surrounds the other, the outer being called the secondary and the inner the primary. Both are shown in what is sometimes called the Dubois-Reymond coil. (Fig. 4.) Much into which we cannot go is to be learned scientifically about a faradic instrument. Its current is not derived from a number of cells. No faradic battery needs more than one or at most two good cells.

Flemming's faradic battery is provided with a commutator or polarity-changer, and with a slow and rapid

Fig. 3.

FLEMMING'S FARADIC BATTERY.

rheotome or current-interrupter, scales by which the primary and secondary currents may be graduated to the utmost delicacy or greatest power, and with a special form of galvanic cell. This cell is so made that when not in action the zinc is raised out of it altogether, and the aperture through which it passes covered with a piece of rubber called a hydrostat, making the cell fluid-tight, and saving both the zinc and fluid from the effects of splashing in transportation, or of immersion in case of an upset. By this plan also the cell can be filled to the top and the zinc be made twice the usual length, thus producing a stronger current and lasting a longer time.

68   THE NURSING AND CARE OF

The Dubois-Reymond coil is provided with slow and rapid interrupters, with switch for making connections with the primary and secondary currents, and with screw for regulating the tension of the currents. It is connected with a cell under the table or elsewhere.

Let me now give somewhat explicit and detailed

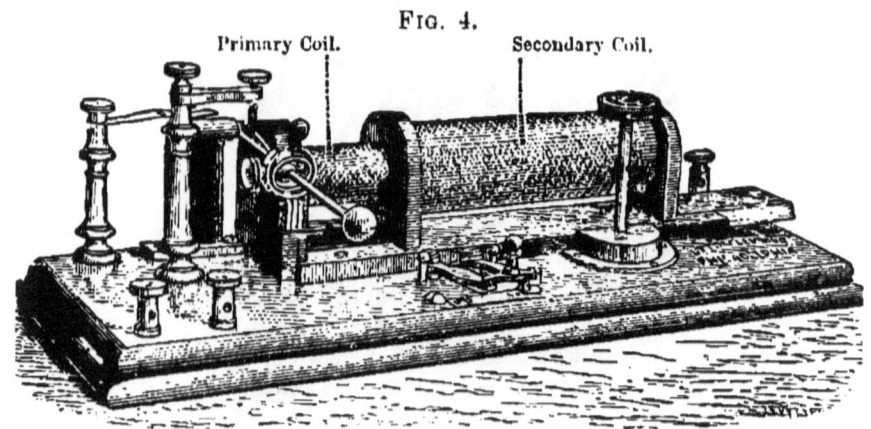

DUBOIS-REYMOND COIL.   IMPROVED BY FLEMMING.

directions as to the methods of using a faradic machine. To some these directions may seem almost too detailed and particular, but they are important matters which belong to the province of the nurse.

To start a faradic current is a simple matter, and yet I have seen it bother a nurse or untrained doctor. Batteries by different manufacturers are started in somewhat different ways, but all are on the same general principle. Understanding the method of starting one, the nurse will probably have no trouble with another. The cell must be in good working order. Many faradic machines have a sliding rod with a joint attached to the zinc, by which rod, when not in use, the zinc may be drawn up, and then kept from slipping back by bending the rod at the joint. The objection to this arrangement is that it necessitates a small zinc which will not endure well. In the Flemming battery, as already stated, the zinc is entirely removed from the cell when it is not in use.

The apparatus shown in Fig. 3 is supplied with a series of switches. One of these is used to complete the circuit through the coils and cells, and also as a commutator or pole-changer. This switch if swung to the right strikes one post, if to the left another; it must be in contact with one or the other.

On the faradic apparatus for office-table (Fig. 5) I will explain what is necessary about the rheotome or current-interrupter and some other matters. The old batteries had only one form of interrupter, a spring lever with one contact screw. This spring gave only rapid interruptions. Some new instruments are made in this way, but this is a mistake. A good battery should have either two rheotomes or a double rheotome, as has the apparatus shown in Fig. 5, one for slow and the other for rapid interruptions. The rapid interruptions are made by an ordinary vibrating spring lever; the slow by a long lever swung horizontally and adjustable by an inclining ring, which regulates the lever's range of vibration. The contact screw is the same for both levers, which are at right angles to each other. It is mounted so as to move horizontally across the angle between the two springs, and can be rapidly swung from one to the other by a simple movement. The method of changing from slow to rapid interruptions is worth knowing. If the contact post is moved carelessly it is apt to scratch the spring. The operator should learn to

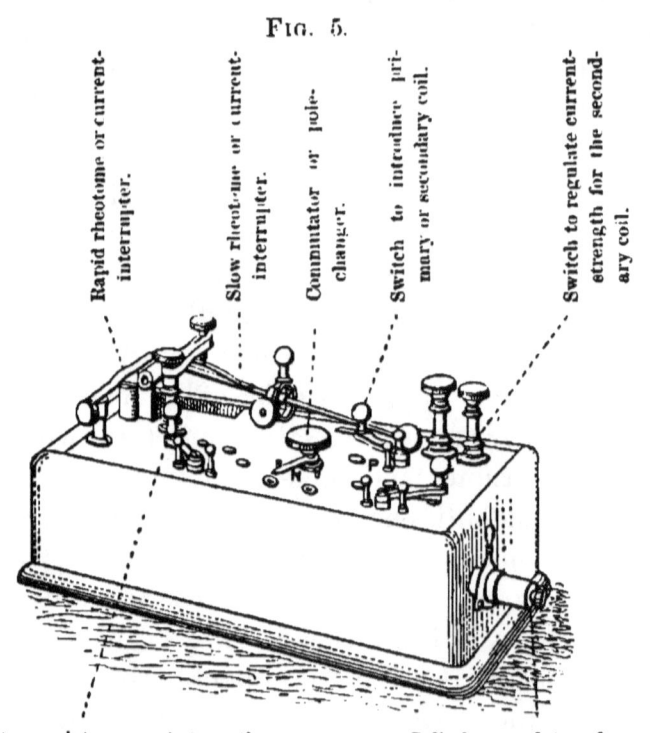

FARADIC APPARATUS FOR OFFICE TABLE.

make this change with one hand, holding down the two springs with two fingers and moving the bar with another or the thumb.

If a very rapid interruption is desired, the contact post is brought over the elastic spring (rapid rheotome,

Fig. 5) and then gently screwed down until it gets to the finest point at which can be had any interruption at all. If it is screwed down too far, the circuit is completed so as not to allow any interruption. If a moderately rapid interruption is required, the screw is withdrawn to a more and more remote point until it is no longer possible for any current to pass. If a slowly-interrupted current is required, the post is swung to the long lever (slow rheotome, Fig. 5). The ring surrounding this lever may be placed at any angle. When placed at a right angle, the result is that the lever swings through the least possible arc, and the slowest possible interruption is obtained. If the ring is adjusted at a somewhat acute angle, the lever is no longer able to move in a large arc, and the rapidity of the interruptions is increased. In this way can be obtained interruptions from one per second up to hundreds in the same time. Batteries are not all made like the one here shown. In one battery the lever moves horizontally and has a somewhat different arrangement; in another an inverted cone within a fork constitutes the slow interrupter. They are all on the same principle,—that is, that of the vibration of a lever in a greater or smaller arc. The rheotome or interrupter sometimes gets out of order. The elasticity of the spring is sometimes impaired by bad usage, or an accumulation of dust on the spring will tend to make an irregular current. Sometimes the point of the contact screw is worn off and needs replacing.

How to "regulate"—that is, how to increase or decrease the current strength—is a most important matter.

Machines from different manufacturers have apparently different methods of regulating the current, but they are all on similar principles. The simplest current-modifier consists of a copper tube sliding over, or covering to any desirable length, the soft-iron core inside the primary coil. It acts under the theory that when a current traverses the primary coil and magnetism is induced in the soft-iron core, a current is also induced at the same time in the copper tube, which, being a closed body and under the influence of the current, acts as a closed coil. When, therefore, the entire length of the copper tube covers the iron core, the current induced in the copper tube counteracts and neutralizes to a certain extent that elicited in the primary coil; and because the secondary coil depends on the inductive force of the primary, also the secondary current. The current is, therefore, weakest when the copper tube covers the entire core, because it checks, by induction in the opposite direction, the primary current. The more the tube is withdrawn the larger is the part of the coil that is made free, the current becoming stronger and stronger until the coil is freed entirely from the contrainduction of the tube. This mode of regulating the current, however, is not entirely satisfactory for the demands of a good faradic coil for therapeutic purposes, since the current, even with the tube-modifier in all the way, is often found too strong for application to sensitive nerves. To meet such cases, the entire coil is divided up into sections, or tapped at different lengths. First-class batteries, such as Flemming's, are provided with switch arrangements, by means of which any one section or any number of sections may be used at pleasure, thus opening currents hardly perceptible by switching into the circuit a small portion of the coil, or increasing the strength by adding coils.

Another mode of regulating the current-strength is found in the DuBois-Reymond induction coil, which, in the absence of a copper tube, is acted upon in this regard by its own secondary coil. The primary coil with the iron core is stationary, while the secondary coil is made to slide over it. When the secondary coil covers the entire length of the primary, the primary is weakest, since both wire-ends of the secondary coil are metallically connected together, this coil becoming one closed body, and behaving precisely like the copper tube enclosing the iron core, counteracting induction. By withdrawing the secondary coil, thus releasing a larger area of the primary coil, the current increases. The reverse action takes place in using the secondary current. The wire connecting the two coil-ends is removed, and the more the secondary coil is pushed in over the primary the more inductive surface is brought to bear, and the secondary current increases in proportion.

Another contrivance for diminishing coarse currents is the so-called water-rheostat,—a glass tube with metallic bottom and top-cap in which is a movable rod. When the tube is filled with water and one terminal post of the battery is joined to the metallic bottom, and the top and the other terminal of the battery connected with the electrodes, the current is forced through the column of water, and increases in strength by gradually pushing the movable rod downwards towards the bottom.

In some older forms of batteries, the iron core is made movable and is used as a modifier; in these batteries the current is weakest with the core out, and by gradually feeding the core in, thus putting more of its iron mass under the influence of induction, the current becomes stronger.

If the battery is a good one, the operator can begin with the weakest possible current and increase to the strength necessary without danger. This should be done by using the two methods of regulating the strength of the current, in an apparatus in which the tube and switch methods are both possible. Suppose with a certain patient it is necessary to be very careful about the use of the battery. The patient may be timid. Some patients imagine that when the battery is applied they are going to have a thunderbolt hurled at them. The best way to begin in such a case is to increase the current by the use of the hollow cylinder. This is pulled out slowly, and, when a certain strength of current is reached, it is left at that point. If necessary to withdraw the cylinder further in order to get sufficient strength, it is well after drawing it out a certain distance to return it and push the switch up one point. The cylinder can then again be withdrawn if necessary; and later, if it is desired, the cylinder can be once more returned and the switch moved to another knob.

Let me now give a hint or two about handling and taking care of a faradic battery. One that is well cared for should not often get out of order. With fresh fluid in the cell every three or four weeks, if the battery is used moderately—say one or two hours every day—it should keep in good working order for a long time. If, however, care is not exercised about some little things it will soon be out of order. The zinc should never be left in the liquid longer than is necessary. A battery weakens quickly if the zinc is not instantly removed. The longer the zinc is in contact with the fluid the longer the circuit is maintained, and the greater is the damage done to the zinc. In the Flemming apparatus, as in some others, a rubber cover is over the opening through which the zinc is removed from the cells. While the zinc is in the fluid certain actions are taking place; hydrogen is generated, and various chemical combinations occur. It is some advantage to leave the cover off for a few minutes, allowing the chemical actions to complete themselves and the gases to escape. After a battery has been used for some time it may have one or two weeks' service in it, and yet it may work lamely. One cause of this is that, after having been in use for some time, hydrogen, which is given off, collects on one of the plates and acts as an element, setting up what are called polarizing currents. It is not necessary to know much about these currents, but it should be understood that by shaking the cell gently this hydrogen will be disengaged and the battery will do better.

The nurse should be able, when the battery is not working properly, to tell whether the defect is in the apparatus, in the current-carriers, or in the electrodes. All that she needs to do is to moisten the fingers and place them at the points where the rheophores are connected with the instrument. If the battery is all right the current will be felt passing through the fingers. The trouble must then be in the cords or in the electrodes. The difficulty is frequently to be found in the cords.

When the apparatus is not being used the switches should be set back to the lowest possible strength,—the first knob of the semicircles. This is an important rule, and one which, I am sorry to say, is frequently violated in hospitals and elsewhere. Remember not only to put the switches back, but also to push in the cylinder-regulator. Keep the apparatus clean in every respect. Remove all dirt. If anything forms through corrosion it should be removed. In removing the zinc from the cells a little acid liquid usually clings to it. The precaution should be taken to shake the zinc and remove as much as possible of the liquid before replacing the zinc in the rubber cup provided for it. The zinc should be kept well amalgamated. When the battery fluid[1]

---

[1] DIRECTIONS FOR USING THE BATTERY FLUID.—Dissolve one and a half (1½) ounces avoirdupois of bichromate of potassium in twenty-four (24) fluidounces of hot water in an earthenware vessel, and add to it three-fourths of an ounce of saltpetre. Allow it to cool to the temperature of the air, and then add three (3) fluidounces of commercial sulphuric acid. When cold again, add a solution of one-fourth (¼) ounce of bisulphate of mercury in three (3) fluidounces of cold water. This will yield one quart of the battery fluid, and should not be used until cold.

DIRECTIONS FOR MAKING THE AMALGAMATING SOLUTION.—Mix half a pound of nitric with one pound of muriatic acid, to which add four ounces of mercury. When the mercury is dissolved add a pound and a half (1½) more of muriatic acid.

DIRECTIONS FOR AMALGAMATING THE ZINC.—Immerse it in

becomes greenish-black and the current too weak to be of any practical service, the fluid should be emptied out and new supplied. Care should be taken where and how this fluid is emptied. It is acid, and might ruin the eyes if it should splash into them. It will also ruin the clothing, or possibly injure the plumbing.

Harm may be done with the faradic battery, chiefly by the use of currents of too great strength. Without going into scientific details with reference to this point, it is sufficient to say that while nerve and muscle can both be strengthened and restored to usefulness through the help of a faradic current, on the other hand, if such current is used of too great strength or for too long a time at one sitting, nerve and muscle may be exhausted by the enforced overwork, and in some cases wasting or other serious trouble may result. It is, therefore, the duty of a nurse to carefully and explicitly attend to the directions which are given to her with reference to the strength of the current and length of the application. The temptation with one not thoroughly posted is to rather overdo than underdo the application ; and patients themselves sometimes assist in bringing about this result by their natural desire to get the most out of the agent that is being used for their relief or cure. Neither single muscles, groups of muscles, nor the limbs should be subjected to currents which are sufficient to twist and contort them so that they are beyond voluntary

---

the solution for a few seconds, then remove it quickly to a basin of clean water and rub it with a brush or cloth, when it will be covered with a fine even coat of mercury. The solution should be kept in a covered vessel, and may be used many times.

control. If a faradic current is applied to the head, before it is of sufficient strength to do harm the pain produced will be so great that, as a rule, patients, if able to indicate their wishes at all, will not allow its further use. Owing to mismanagement of the battery or of the cords or electrodes, however, a very strong current, possibly the full force of a coil, may accidentally be applied to a patient's head or neck, or other part of the body, and cause a serious jar, contortion, or other disturbance. Patients who are unfamiliar with electricity should always be handled with gentleness, and if strong applications are required these should be gradually led up to by the careful use of the graduators or regulators. Many persons have a deep-seated dislike, or even fear, of electrical applications, and this is not best overcome by beginning their treatment with a strong shock. The abdomen, next to the head and neck, is perhaps the most susceptible part of the body in making faradic or other applications, and it is often well to moderate the strength of the current in passing from applications to the posterior part of the trunk or limbs to the abdominal muscles.

In a galvanic battery the current which is used is derived directly from the cells. Any two dissimilar substances, as zinc and copper, or zinc and carbon, immersed in acid or saline liquid, and then brought into contact, will give rise to electricity, and, in fact, constitute a galvanic battery of one cell. Four, eight, sixteen, thirty, or any number of cells may be combined in a galvanic battery, as is found in practice. Various contrivances for starting, measuring, interrupting, modifying, or changing the polarity of the electricity are usually attached to the galvanic instruments of physicians, but these have nothing to do with the origination or calling forth of the force, true galvanism being the result simply of the contact of the unlike bodies and the action going on in the cells. I make these remarks because one unfamiliar with the subject, seeing coils, switches, etc., on the key-board of a galvanic battery, may think that these have something to do with the origination of the current, which is not the case.

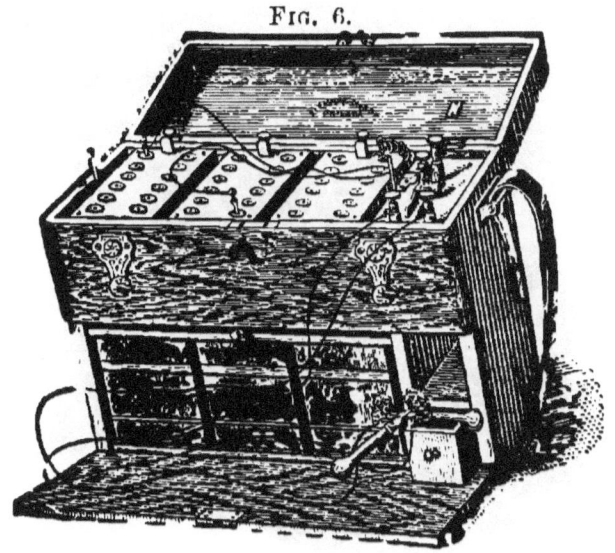

FIG. 6.

PORTABLE CONSTANT GALVANIC CURRENT BATTERY.

In Fig. 6 is shown a galvanic battery of thirty cells made by Flemming. In this battery if the operator desires to use the current of only a small number of cells, he need put but a single section in action, thus

80   *THE NURSING AND CARE OF*

saving the zinc and fluid of the rest; or if any accident should happen to a part it affects only the section it is in, and not, as in some other forms, the whole battery. The elements are zinc and carbon, the fluid the bichro-

Fig. 7.

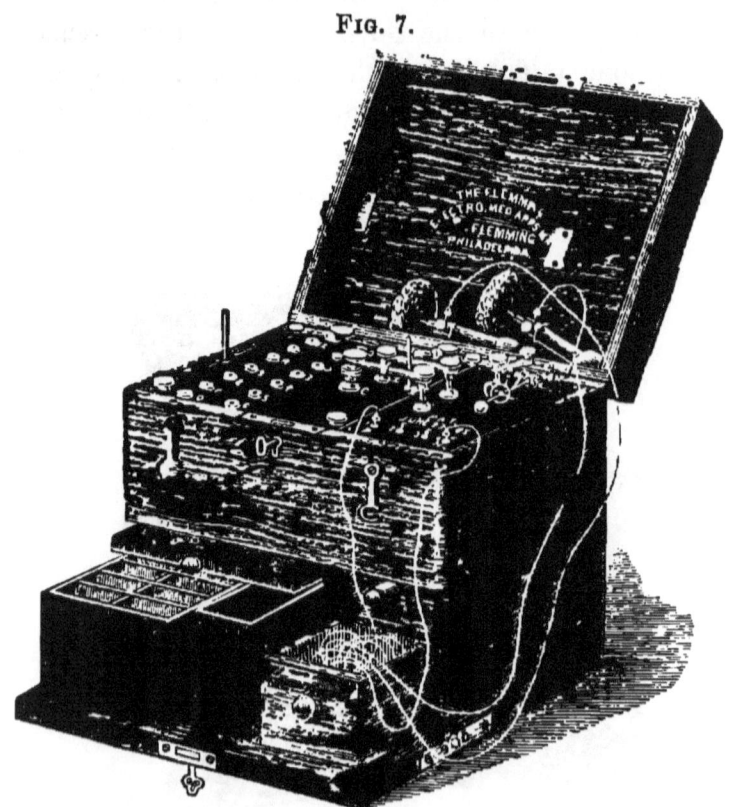

FLEMMING'S 30-CELL COMBINATION BATTERY.

mate of potassium solution, and the cells hard rubber, which is more durable than glass. The battery is put into action by raising the sections of the cells by rods at the back, these rods being so regulated by springs that the elements can be immersed to any desired depth, thus regulating the quantity of electricity evolved at pleasure. Covering the cells when not in action is a rubber hydrostat, or rubber-cushioned sliding-board, which, by means of two small rods on either end, is pressed down tightly upon them when the lid is closed, thus preventing the fluid from spilling during transportation, or if by accident the battery should be upset. Unlike many other batteries, the cells of this can be taken out in front for recharging, obviating the trouble and the risk of breakage in removing the elements. Connected with the binding posts for receiving the electrode cords is a commutator for reversing the polarity of the current or interrupting it by hand.

In some instruments the galvanic and faradic apparatus are combined (Fig. 7), although, as a rule, I believe it better to have them as separate instruments. In application, either the slow or rapid interrupter of the faradic apparatus can, by simply moving a switch, be made to act as an automatic rheotome for interrupting the galvanic current.

In Fig. 8 is shown a galvanic battery of no less than fifty cells, weighing but eleven or twelve pounds, and only twelve and a half inches long. It is manufactured by John A. Barrett, 13 Park Row, New York, and, as it is a compact and convenient instrument, the nurse may sometimes be called on to use it. The fifty cells are placed in a box, and a top plate, perforated for the passage of posts or studs from each cell, is laid over it. In the lid of the box is placed a commutator or pole-changer, to which are attached short, flexible cords,

f

Fig. 8.

BARRETT'S CHLORIDE OF SILVER BATTERY.

terminating in hollow plugs. By these cords and plugs any number of cells may be selected from any part of the battery. The pole-changer is also provided with two binding posts, for the conducting cords.

These are portable galvanic batteries. They are used by connecting the rheophores or current-carriers directly with a certain number of cells and then with the patient. No coil is interposed between the cells and the patient, although often, as I have already stated, some forms of galvanic battery have certain accessory appliances which are misleading to a person who is not familiar with the subject.

Some manufacturers represent that from a faradic apparatus, consisting of only one cell with its accompanying coils, a galvanic and also various forms of the faradic current can be derived. It is true, theoretically, that a galvanic current can be obtained from one or two cells, but in practice such a current is of no use, and, therefore, the claim is an unjust one. Those who have examined such instruments know that the so-called galvanic current is nearly always simply a very weak faradic current.

Some batteries are put into use by raising the cells upon the elements (Figs. 6 and 7), and others by dropping the elements into the cells. In others the elements remain permanently fixed in the cells, as in the Barrett battery (Fig. 8). Most permanent office batteries are made in this way. In some batteries all the cells are lifted at once, but this is a wasteful method.

The regulation of the current-strength depends upon the number and character of the cells in the circuit. In the battery shown in Fig. 6, two cords with metallic points are attached to the same binding post. By placing the pins at different points from two to thirty cells can be used. The second cord is a contrivance to avoid giving unnecessary shocks to the patient while increasing the current-strength. If, for instance, the nurse is applying fifteen cells and wishes to increase to twenty, instead of taking out the first pin and placing it in No. 20, she allows the first to remain until she puts the second in No. 20; because when the current-strength is increased or decreased, and at the making and breaking of a circuit, a new current is induced in the patient, and it is this which causes the shock. A practical point, then, in the management of the battery is to remember to always be careful how to begin and end an application, or how to increase or decrease the strength of the current.

The commutator or pole-changer is a necessary attachment to every galvanic battery. This is a contrivance which not only makes and breaks the current, but with it the polarity can be changed from moment to moment,—that is, it can make the binding post, which is the anode or positive pole, become the cathode or negative pole without removing the electrode from the patient. This commutator or pole-changer is explained very well in Dr. Amidon's "Student's Manual of Electro-Therapeutics;"[1] from which I have borrowed by permission the diagram and the following explanation.

"The accompanying cut will show that this changing of polarity is accomplished by means of a split button, which consists of a disk of hard rubber revolving on a shaft which is bound with a tire of brass in which two breaks occur. Four springs (s' s'' s''' s'''') press firmly against this tire in such a way that two of them, when the button is rotated, jump over the gaps and come in contact with the opposite half of the tire.

"For instance, in Fig. 9 the current leaves the pos-

---

[1] "Student's Manual of Electro-Therapeutics," by R. W. Amidon, M.D. G. P. Putnam's Sons, New York, 1884. This is a small but valuable book on medical electricity, especially well adapted for students or for nurses who wish to obtain a practical knowledge of electricity without going too deeply into the subject.

itive pole, traverses the spring s'', runs around the right half of the tire t' through the spring s'''' to the binding post 1. Then, if the circuit be closed, across to post 2, back to spring s''''', around the left half of the brass tire t'' through the spring s''' to the negative

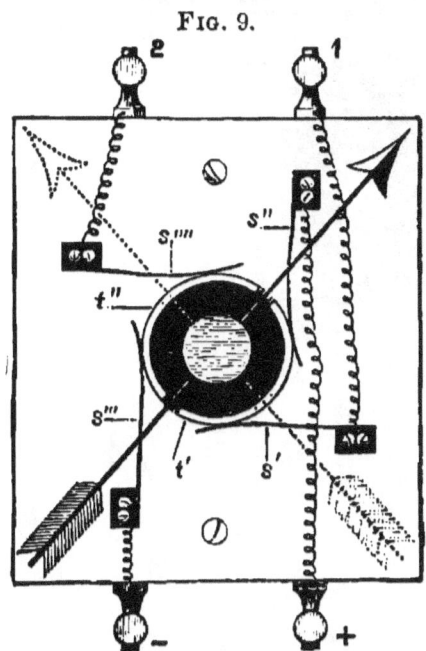

Fig. 9.

DIAGRAM OF COMMUTATOR.

pole. This will make 1 the positive pole, 2 the negative pole. If the button is turned to occupy the position indicated on the dotted arrow, it is easy to see the polarity of the posts will be reversed,—*i.e.*, the current will run from the + directly to 2, and from 1 straight back to —.

"This split bottom is covered in most batteries, and is turned by a handle or winch, which should always point as does the arrow in Fig. 9,—towards the positive pole. There should be two binding posts, which should be made to hold the rheophores or conductors by means of screws."

Fig. 10.

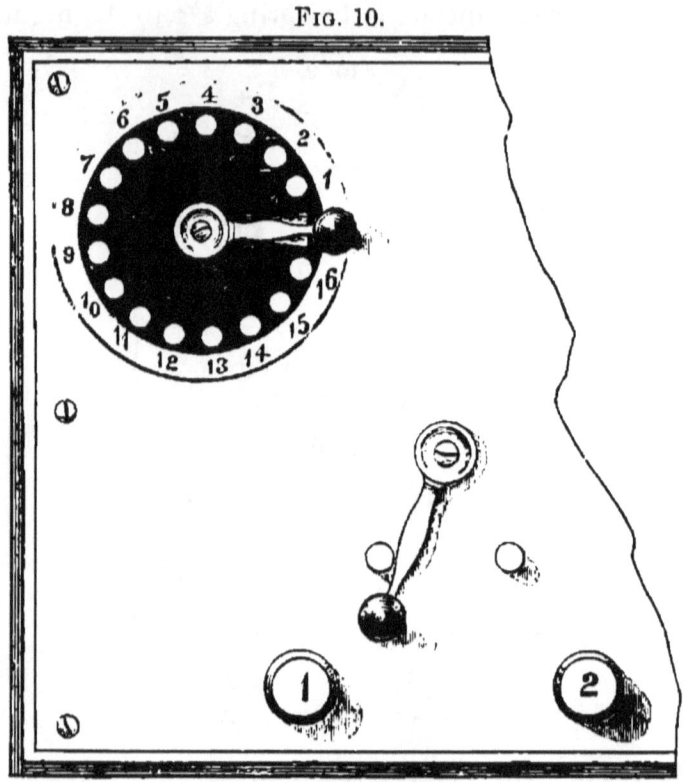

COMMUTATOR SWITCH.

The nurse should have always in mind certain points about the care of the battery. She should be careful about spilling the fluid of the battery, for this is acid. There is sometimes some difficulty in letting the cells down or lifting them. Under such circumstances nothing can be accomplished by force. If rough, the operator will be likely to break or spill something. With a little careful manipulation the rod slips down without any difficulty.

On the whole, I think it is better for nurses never to use the galvanic current to the head or neck. Certainly they should not do this unless the instructions which they have received have been very explicit, and with reference to the particular case which is being treated. When it is recalled that a great French physician, one of the founders and best exponents of electro-therapeutics, instantaneously destroyed the sight of one of his patients by an incautious galvanic application, the necessity, aye, the great importance, of both physicians and nurses knowing the dangers of the agent with which they are dealing is very manifest. The manner in which this accident occurred indicates one of the dangers, and suggests the precaution which should be taken in order to avoid such an untoward event. While a galvanic application was being made to the head, the current was suddenly increased to one of many cells. A great sheet of light appeared before the eyes of the patient, and the sight of one eye was lost forever. This event was doubtless brought about by the complete exhaustion of the retina by the sudden, terrific overstimulation to which it was subjected. If the muscles and the coarse nerves of the trunk and limbs can be exhausted by electricity, much more certainly can the delicate retina be exhausted, and its function impaired or destroyed by this agent. Doubtless, also, the nerves of hearing, smell, and taste can likewise be injured by electrical over-stimulation, although perhaps the danger to them is not quite so great. Indeed, I now recall that an enthusiastic young German scientist, Ritter, through experiments performed upon his auditory nerves for scientific purposes, destroyed or impaired his sense of hearing. Vertigo can be and frequently is produced by electrical applications to either the head or neck. Great differences in individual susceptibility are noted in this respect. Vertigo will be produced in some by very weak currents; in others currents of considerable strength can be well borne. Incautious applications may go further than the mere production of vertigo; syncope or fainting may result, as I have myself seen. Just how these results are brought about is a question of theory. It is probably through the direct or reflex effect upon the pneumogastric nerve. The practical point for the nurse to know is that such dangers exist. Knowing this, she will be willing to be guided by the physician, and will not be willing to take any improper risks on her own responsibility.

Ordinarily the doctor directs the use of so many cells—fifteen, twenty, thirty, etc.—of a galvanic battery. This is not an exact standard. The current-strength given by ten cells on one day will differ from that obtained on another. It depends upon many varying circumstances; therefore when applications are gauged by cells the method is very inexact. A recently-invented instrument enables the doctor to order the nurse to give a certain dose, and it enables the doctor or the nurse to record the dose given. It is called a milliampèremetre. (Fig. 11.) All that can ordinarily be seen of it is a needle, which is made to move across the arc of a circle.

This is arranged from zero in the centre to fifty on each side. These spaces measure so many milliampères. A milliampère is resistance offered by a human body to a

Fig. 11.

THE MILLIAMPÈREMETRE.

current of electricity which is generated by three Daniel's elements,—that is, by three cells of a certain kind of battery. It is a term introduced by Dr. de Watteville, of London.

Dr. Landon Carter Gray,[1] of Brooklyn, describes the milliampèremetre made by John A. Barrett, which is shown in Fig. 11, as follows:

"The scale is graduated from one to fifty of these new units. The working parts of the metre are enclosed in a case about four inches square at the base, and six inches high, having a glass face showing pointer and scale. The pointer is attached to a movable magnetic

---

[1] *New York Medical Journal*, July, 1885.

needle, which is a steel disk suspended between a surrounding coil of wire on two knife-edges, so as to oscillate freely in a vertical plane. The movements of the magnet are indicated by the pointer on the scale. The magnet is counterbalanced and held in equilibrium, so that the pointer rests at zero, by means of a small weight fixed to it below the centre of gravity. These are so arranged that the position of the instrument relatively to the earth's magnetic poles is of no account. The magnet, being so poised on frictionless knife-edges, would be subject to prolonged and troublesome oscillations; but a dampening device is provided, which consists of a vane or fan of thin mica, extended upward from the magnet in the back part of the case by means of a light, rigid stem of aluminium. This vane moves broadside against the confined air and brings the needle quickly to rest in any position without, in the smallest degree, interfering with its sensitiveness. A locking arrangement also forms part of the instrument, by which the magnet may be lifted from its bearings and fixed in an immovable position. By this means the knife-edges and movable parts are protected from damage during transportation and when not in use. A levelling screw is placed on one side below, so that the instrument may be readily adjusted to a proper position."

The instrument is put into the circuit with the patient. A galvanic battery, for instance, is taken and nine or ten cells put into the circuit; then a cord is run from one binding post of the battery to one post of the milliampèremetre, while another cord is connected with the other pole of the battery and goes to the electrode. The other electrode is connected by a conducting cord to the remaining post of the milliampèremetre. The electrodes are then placed upon the patient.

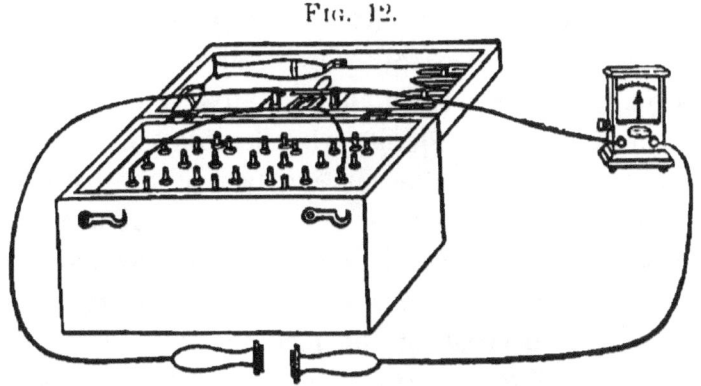

FIG. 12.

DIAGRAM SHOWING THE METHOD OF USING THE MILLIAMPÈRE-METRE.

Watching the milliampèremetre when a current is passing, it will be seen that the needle swings to a certain point, where it remains a certain length of time. That point is noted. To produce the contraction of a certain muscle, it may to-day require three milliampères; perhaps on another occasion it will need four and a half or five, and thus the operator is enabled to make an accurate record. To-day ten cells may give three milliampères, while to-morrow twice as many may be required.

While I make these remarks about the usefulness of the milliampèremetre, it is important to acknowledge that up to the present few if any absolutely correct instruments of this kind have been made. They are being improved, and we will probably eventually have one that will be both convenient and accurate. It should be remembered that all milliampèremetres require to be handled with great care, as from rough usage they soon get out of order.

The term electrode or pole is applied to the instrument with which electricity is applied to the patient. The term rheophore or current-carrier is sometimes used to indicate the same appliance, but it is more properly employed to designate the flexible conductors or cords which go from the binding post of the battery to the electrodes or handles. I will here confine myself to a few practical general remarks about electrodes and rheophores, but may say more about them when I come a little later to speak of the methods of applying electricity to different portions of the body. Electrodes are large or small; they are of different shapes, and are designed for making application to various organs and parts,—to the skin, muscles, nerves, joints, internal organs, etc. As a rule nurses will not be called upon to use the more special forms of electrodes for internal applications, as to the throat, bladder, urethra, or uterus. It is beyond their province to make such applications; therefore, it is not worth while to study the forms of electrodes used for these purposes. They will be concerned only with the use of the commoner forms of electrodes. These are usually disks of metal or carbon connected by a screw with wooden, vulcanite, or other insulating handles. They are also, of course, joined to the conducting cords which run to the battery. The nurse should know how to select the best form of electrode for accomplishing certain purposes under the direction of the physician. When, for instance, an application is to be made beneath a patient's clothing, a good form of electrode is one in which the disk is in the same plane as the handle. In most electrodes the plane of the disk is at right angles to that of the handle, and this makes it inconvenient to slip beneath the clothing. An interrupting handle is sometimes to be used, and should be understood by the nurse. Various forms of interrupting handles are made, but they all consist essentially of a contrivance by means of which the current can be opened or closed by the operator at any moment without removing the electrodes. The simplest and commonest form is one in which, by simply pressing down a spring, the circuit can be made, and by stopping the pressure can be broken. Skill in the use of an interrupting handle is obtained by a little practice. Instead of an interrupting electrode to be held in the hand, sometimes a foot or pedal interrupter is used, but when employed the physician should specially explain its method of working.

Never use a sponge or moistened electrode of any kind which is not clean. It ought to be unnecessary to say this, but it is a rule that is sometimes violated. If the nurse is making a series of applications to the same patient, it may not be objectionable to use the same electrodes continuously; but even in such a case the coverings should be cleaned after each application. They should be washed thoroughly in warm water, or with some form of corrosive sublimate or carbolic soap, or other good cleansing material. After a time electrodes may become incrusted from being dipped into salt solutions, or certain chemical deposits may take place on them. These deposits on the surface of the electrodes may cause local electrical actions, and render the application unnecessarily painful. A clean, polished electrode is, therefore, less painful than any other.

A neat method of covering an electrode, suggested by Dr. G. B. Massey, of Philadelphia, is to employ

Fig. 13.

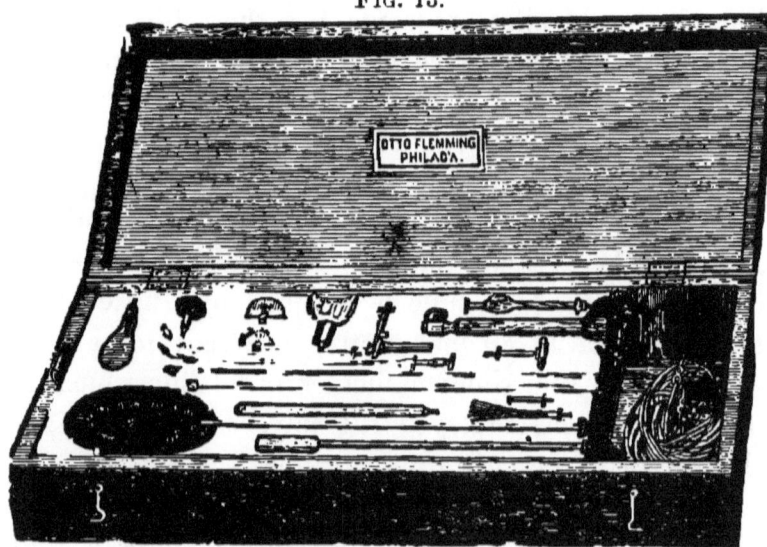

BOX OF ELECTRODES.

absorbent cotton instead of chamois skin or sponge. Take a piece of clean absorbent cotton, fold and twist it over the disk of the electrode, and wet it. It will cling to the disk, and after the application it may be thrown away and a fresh piece used every time.

Physicians and nurses should learn to hold the electrodes in an easy and advantageous manner; if possible, both should be held in one hand. This is important for various reasons. It leaves the other hand free to regulate the strength of the current and for other purposes. In direct muscular applications, and in all others in which it is possible to hold electrodes in one hand, this should be done in a certain way. They should be held, pointing backwards, the handle of one electrode between the index and middle finger, and that of the other between the ring and little finger, as shown in Fig. 14.

Fig. 14.

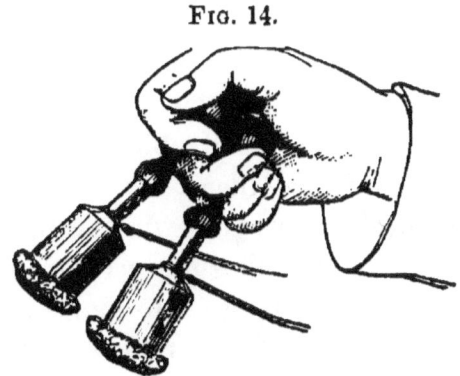

METHOD OF HOLDING ELECTRODES IN ONE HAND.

Held in this way, the disks can be brought very close together, or can be separated several inches. Almost the entire length of the forearm or upper arm can be included between the electrodes thus held in one hand, or by shifting one of them between the thumb and index finger. The nurse should practise employing one hand wherever it is possible.

It is not unimportant to know something about the connecting cords which are used in making electric applications. A good conducting cord should have certain qualities. It should be flexible, not easily broken, and capable of being easily repaired. The operator should know how to detect any break or defect in the cord. Most conducting cords are made either of tinsel or of more or less flexible copper wire. If it is found that the current is not being conveyed properly the cords should be looked at, particularly at the points where the wires are attached to the terminal pins. The best cord is, I think, a somewhat flexible twisted copper wire, insulated and attached to the pins in a peculiar manner by double eyelets. The practical thing to do is to have two sets of cords,—one of insulated flexible wire, the other several yards of "cable wire." I generally have three or four yards of this "cable wire" in the drawer of the battery box, and then I am not at a loss if the cords should break. This cable wire is copper wire about No. 16 or 17, covered with gutta-percha. If it breaks, all that it is necessary to do is to scrape off the gutta-percha and repair it. To shorten it, it can be broken at any point.

In taking up the methods of applying electricity I shall restrict myself to those which are in most common use,—those which should alone be employed by the nurse under competent directions. I shall discuss applications to the muscles, nerves, skin, joints, limbs, and to the spinal cord, and certain general methods of treatment. Nurses will chiefly have to deal with neuro-muscular applications,— applications to nerves and muscles.

To make a deep-seated application to a muscle or nerve the electrode should be moistened or wet. It was discovered by Duchenne—although it seems so small a matter that it can hardly be spoken of now as a discovery—that an application can be restricted or limited to the skin by the dry method, and that by the wet method the skin can be almost entirely neglected and the deeper-seated parts reached.

In making applications to muscles or to nerves supplying muscles, in cases of paralysis, two different methods, known as the direct and the indirect methods, or direct muscular electrization and indirect muscular electrization, are commonly used. In direct applications the electrodes are applied as directly as possible to the muscles. In the indirect method the muscle is caused to contract through the nerve or nerves which supply it. In order to understand the indirect method it is necessary to know something about motor points. These are points where the nerves which supply the muscles are most readily reached by the electrodes. They are points where the nerve bends round the muscle, or dips beneath it, or passes into it, or comes out of it.

To a certain extent motor points can be studied readily by any one. Placing one electrode at any convenient place, as on the sternum or along the spine, and then gliding over any selected muscle, a point is found where it contracts most readily. This is the motor point for this muscle. Suppose we have a case of so-called wrist-drop; the hands hanging limp from want of power in some of the muscles of the forearm. The electrode is placed at one point and the hand moved inwards; at another it moves outwards, but the movement desired is upward, and finally a point is found where extension or lifting of the hand is produced. Certain movements belong to the foot,—upward or downward flexion or extension, and certain combinations of directions, as outwards and upwards, inwards and upwards, inwards and downwards, etc. It is necessary to know in a general way where the muscles producing these movements are situated; then, if a certain form of paralysis is to be treated, the operator will soon learn to what points the electrodes should be applied in order to do the most good. Something should also be known of the muscles producing the general movements of the limbs and trunk, and how to cause these movements by electricity.

I will give a few hints and cautions about muscular applications. In the first place, the operator should be satisfied to make a moderate application, and thus avoid one of the mistakes often made by the nurse and sometimes by the physician. If the contraction is distinctly visible, it is sufficient. The patient will often urge the nurse or physician to use more, on the principle that if a little will do good a larger quantity will do more good. The limbs should not be violently contorted, nor should the applications be continued too long. General faradization, to be described later, if done skilfully, either by the lying or the sitting method, can be completed in from twenty to forty minutes. The nurse simply passes rapidly from one muscle to another, or one muscular group to another, producing contractions for a few seconds. If the application is continued too long, or made too severe, it may cause exhaustion and even wasting. Although these patients may improve for a time, the benefit will not be lasting.

One reason why electricity is often unsuccessful is because doctors and nurses are slipshod in its use. It is not always an interesting matter to apply electricity, but any one who does not intend to do it properly should not attempt it at all.

Electricity should never be used soon after meals, for electrical applications may interfere with digestion. The preferable time is midway between meals, or as far away from meals as possible.

Another practical point is that electricity should never be applied to a limb or muscle which is in a tense condition. The muscles should always be relaxed, but the opposite of this is frequently the case. The patient will stretch the arm to its full length, or will stand while electricity is applied to the back of the leg and calf. This is improper. If the patient is standing at all, the application should be made to one leg while the weight is borne on the other; it is better to have the patient lying down. So in the arm it is well to have the arm raised half-way, so that the muscles shall be relaxed. The affected arm may be supported by the well hand.

Faradic electricity will do for muscular applications in most cases; it will do in all cases of paralysis due to a central lesion; it will answer in all cases where the muscles respond readily to it. If the affection is due to rheumatism, injury, lead-poisoning, or other so-called peripheral cause, producing what the doctor calls a peripheral paralysis, it will be difficult or impossible to get a response with faradism, and galvanism will have to be used.

In making applications of electricity to the muscles of the face it is important to proceed with great care. The current which is applied should not be too strong, and its strength should be very slowly and cautiously increased. In such application it is necessary to use small motor-point electrodes, as it is impossible to reach the individual muscles unless the electrodes are small. One pole is placed upon an indifferent part, or as near as possible to the point where the nerve emerges from the skull. Nurses should never make these applications except by explicit orders and under observation.

In making an application to the skin, a rapidly-interrupted faradic current is usually preferred by the physician, although a galvanic current may sometimes be used. A metallic electrode should always be used,—unless other directions are received,—a brush, or one of the uncovered disks already shown. One of the electrodes is to be moistened and the other may be a wire brush. The moistened electrode is to be placed at an indifferent spot, as on the sternum, at the lower part of the spinal column, over the knee-pan, or at the upper part of the thigh. The important electrode is the dry one. The surface to be treated with the metallic brush or disk should be gone over rapidly. A large brush electrode should be used where a large extent of the skin is to be treated.

Another way of making applications to the skin, particularly to the skin of the face and head, is by the hand; and this is the only sort of application that the nurse should be allowed to make to the head. One electrode is put into the hand of the patient and the other taken in the hand of the operator, and the free hand of the latter, dry, is then passed over the skin of the face, neck, and head. The back or front of the hand may be used.

If it is desired to make an application to a joint, this may be done in two ways. One is to take two broad, wet electrodes, or one large and one small electrode, and, placing one on one side of the joint and the other on the opposite side, allow the current to pass through the part. The galvanic current is the one generally used for this purpose. Another method, which is useful in applying electricity to the joints of the foot, is to take a basin of warm water to which a little salt has been added, and place the electrode in the water. The foot is immersed in the water and the other electrode applied to some portion of the limb.

For spinal applications the galvanic current is used. The spinal cord cannot be reached, except reflexly, with the faradic current. The patient should lie on one side if in bed. It is not necessary to uncover the patient. One electrode may be slipped under the clothing from below upwards and the other from above downwards. One may be placed over the sacral region, and the other at the upper portion of the spinal column, and held there for a definite time. It is then moved a little further down, step by step, until the whole application of fifteen minutes is completed. This is very tedious; but, to accomplish any good, the electrodes must be held firmly to the spine and moved from one point to the other until the whole spine is included. The physician may give specific directions that an ascending or descending current shall be employed. In using the ascending current, the positive pole, or anode, is placed at the lower portion of the spine and the negative pole, or cathode, above. In applying the descending current, the positive pole is placed above and the negative below. Another method of making spinal applications, which is useful in some cases of neuralgia and in some special spinal troubles, is to apply the electrodes along the spine and out along the spinal nerves. One electrode is placed over the spinal column and the other is glided around one-half of the body, following the lines of the nerves which come off from this portion of the spinal cord. This is continued from the top of the spinal cord to the bottom, or the reverse, first on one side and then on the other.

As to applications to the brain I have nothing to say, because a nurse should never make these applications. A nurse can, when directed, make applications of faradism to the head with the moistened or dry hand in the manner described, but the application of the galvanic current to the brain I have already stated is not devoid of danger in inexperienced hands. No nurse, likewise, without special training, is capable of making applications to the larynx, the uterus, the bladder, or other internal organs which are reached in any other way than directly through the skin. Applications may be made, under directions, to the intestines for constipation, or to the stomach or liver; but applications to the internal organs, which require special manipulations, should be always made by the physician.

Nurses may be called upon to use the faradic current

for general treatment,—that is, to make applications to all, or nearly all, parts of the body. This so-called general faradization may be used upon the patient either in bed or sitting up. In what is commonly termed the "rest treatment"—the treatment by seclusion, rest, massage, and electricity—the electricity is applied to all parts of the body in succession, the patient remaining in bed. Frequently nurses are called upon to make such applications. Well-wetted electrodes are used, either sponges or absorbent cotton, which are dipped in warm water to which in some cases a little salt may be added. The patients for whom the rest treatment is appropriate usually have no changes in the nerves and muscles which would call for any particular selection of the galvanic or faradic current; in other words, the nerves and muscles commonly respond readily. For this reason the faradic current which is most readily managed and applied is most resorted to in these cases. Beginning with the limbs, preferably the lower extremities, the electrodes are applied either directly over the muscles or to the motor points. Care must be taken to include as far as possible every muscle and group of muscles in the application. Contraction is first caused in the muscles which move the toes and foot; next the operator passes to the legs and thighs, taking in each set of muscles in turn,—the posterior, anterior, outer, and inner. Having completed the application to the lower extremities, the upper limbs are next faradized, beginning with the fingers and hands and ascending to the shoulders. The muscles of the back, chest, and abdomen are treated last. Sometimes it is advised to treat the muscles of the back, loins, and abdomen before the upper extremities. Perhaps it makes but little difference as to this matter, but on the whole I believe it is most convenient usually to leave these until the last. The strength of current should be just sufficient to produce visible movement of the muscles. Violent contorting of the muscles should be carefully avoided. From time to time the electrodes should be re-moistened, care being taken always to squeeze out the water thoroughly in order not to have it trickle over the patient. Applications are not made to either the head or the neck unless they are especially directed. In making applications to the shoulders and chest, care should be taken not to carelessly place the electrodes over the side and front of the neck.

The second method of general faradization—that used upon the patient in a sitting position—is a method often employed by physicians in their offices, and may sometimes be used by nurses under direction. Drs. Beard and Rockwell were among the first to make this method popular. According to them, the use of a faradic current for half an hour or less daily has a decidedly tonic effect upon the nervous system. Other observers believe this observation to be correct, hence electricity is frequently used in this way. The special mode of treatment differs a little according to choice or convenience. One method is to place in the hand of the patient one of the well-wetted electrodes; the other is at first held by the operator, who with his free hand makes application to the head, face, and neck. A weak rapidly-interrupted primary current is usually employed. After this careful application to the head, face, and neck has been made, the patient still keeping the electrode in the hand, the operator passes down the back and arm with the other electrode. The patient then shifts the electrode to the opposite hand and the operator passes down the arm of this side. Both electrodes can be taken into the hand of the operator and rapidly passed or glided over the arms and trunk. The lower extremities are treated last. A good plan is to have first one foot and then the other placed upon a flat electrode, or both feet can be placed upon a large metallic electrode covered with moistened cloth or chamois skin; with the other electrode the operator passes carefully over all accessible portions on the lower extremities. Instead of this method it is sometimes more convenient to keep one electrode upon the lower portion of the back, while the other, passed under the clothing from below, is applied to muscles and muscular groups in succession from the feet to the hips. At certain positions, as under the knee, the large nerve-trunks are more accessible than elsewhere, and by occasionally applying the electrode in these positions any muscles which may have been omitted in the direct application can be reached more or less thoroughly, according to the current's strength, by way of the nerve supply. Unnecessary exposure should be carefully avoided. A little ingenuity and forethought will often enable the operator to make his entire application with comparatively little exposure.

CHAPTER IV.

The Nursing and Care of the Insane.

WITH reference to many matters that are sometimes included in the published or oral instructions which are given to nurses and attendants upon the insane, it is not my purpose to deal in the present chapter. In other works upon nursing—surgical and medical treatises in particular—specific directions are given with reference to many things necessary to be known by those who are in attendance upon the injured or sick, whether sane or insane; such as dislocations and fractures; sprains, wounds, and hemorrhages; artificial respiration; the use of compresses, poultices, stupes, and fomentations; the treatment of bed-sores and ulcers.

Very little has been published with reference to the nursing and care of the insane, except what may be found under the head of treatment in the general text-books on insanity. Two small books have appeared within a year or two, one in Great Britain and one in this country. The first, entitled "A Hand-Book for the Instruction of the Attendants on the Insane," was prepared by a sub-committee of the British Medico-Psychological Association, appointed at a branch meeting held in Glasgow on February 21, 1884, and has been published in this country by Cupples, Upham & Co., of Boston. It contains much useful information and valuable instruction, but too much attention is devoted to the consideration of elementary anatomy and physiology. A more valuable work is entitled "How to Care for the Insane. A Manual for Attendants in Insane Asylums." By W. D. Granger, M.D., First Assistant Physician in the Buffalo State Asylum for the Insane, and published by G. P. Putnam's Sons, New York and London. I have obtained from both of these books some valuable facts, hints, and suggestions; but most of my remarks will be based upon observation and experience, and upon information obtained directly from specialists, and from the medical officers and attendants of insane hospitals with which I have been connected as consultant.

While a knowledge of such subjects as the anatomy and physiology of the nervous system and the nature of the different forms of insanity may be of value to nurses and attendants, it is not strictly necessary that they should be fully informed about such matters. There is, indeed, a certain danger that if too much attention is paid in the instruction of nurses and attendants to subjects which belong in strictness to a true medical education, they may make the mistake—a not infrequent one among them—of supposing that they are the doctors. The information absolutely necessary for nurses to have about the structure and functions of the nervous system can be compressed into a very small compass. It is necessary that they should know that the brain—with the details of the treatment of whose diseases they are sometimes intrusted—is an organ which, although deeply seated within a thick and strong bony chamber, is subject, through the multitude of nerves which put it into communication with the outside world, to impressions which soothe, which annoy, which comfort, which distress, or which irritate.

It is well that nurses and attendants upon the insane should have some knowledge as to the forms of insanity and the special symptoms exhibited by certain patients; but, as already intimated, this knowledge need not be very extensive nor profound, though it should be exact as far as it goes. When it is remembered that even educated physicians, unless they have spent some time in the special study of insanity, have considerable difficulty in separating the different forms of insanity, it is easy to be seen that it would be folly for an individual who is not medically educated to try in a little brief instruction to obtain such information. The superficial features of a case will not enable an opinion to be reached as to its nature. Thus in melancholia, which is a form of mental disease, the essential feature of which is emotional depression, a patient sometimes has the violent outbreaks of agitation or excitement which are really the result of depression; and, conversely, a patient suffering from true mania may, for brief intervals, be in a state of great moodiness and depression. We might illustrate by references to other well-known forms. What a nurse should know is that there are certain great divisions of diseases of the mind which give special symptoms, and that patients must be cared for in this way or in that according as they suffer from one or the other of these forms.

The doctor having once, in the presence of the nurse or attendant, clearly made the diagnosis of the form of mental trouble, the latter, knowing this, should bear certain facts in mind: should remember, for instance, that a case of melancholia is likely to be suicidal or to starve to death; that one of mania is not infrequently homicidal or destructive; that one of monomania may exercise duplicity and may suppress for a time his delusions; that a dement is likely to be filthy and not to take care of himself in any way; that an insane epileptic may one minute be peaceful and serene and the next may be in a most dangerous or motor maniacal paroxysm; that the hysterical insane may make false or pseudo-attempts at suicide. To illustrate still further, the nurse or attendant should know that, for various reasons, accidents are likely to occur among cases of paretic dementia,—the "general paretics," as they are commonly termed in the hospital. These patients, although often exceedingly weak and frail physically, owing to their peculiar mental condition, to their delusions of strength and grandeur, are likely to be demonstrative and pugnacious. Frequently they have weak hearts, or degenerated livers or kidneys; often, as in locomotor ataxia,—a disease closely allied in pathology and often merging into it,—their bones are very brittle; not infrequently they have lung troubles. If such patients are handled as roughly as a case of ordinary acute mania, the consequences may be serious. Collapse has been known to come on almost instantly after a struggle with a patient of this kind, and in badly-regulated asylums bones are frequently broken; these accidents occurring even when no cruelty or unusual violence was intended or attempted. Attendants should remember that, as a rule, these patients are not as strong as their demonstrations would make them appear. A serious accident, resulting in the death of the patient and the subsequent trial of the attendant on a charge of manslaughter, occurred not long ago in one of our large hospitals, the patient being a general paretic with pugnacious and aggressive symptoms, but with advanced mental and physical degeneration. It is in just such matters as flow out of considerations of this kind that attendants require to be informed; they need that kind of knowledge of mental disease which will enable them to discriminate in their management of patients of different classes. What is known as hæmatoma auris, or the "insane ear," is not infrequently present in general paretics, and is another evidence of degeneration. The external ear becomes enormously swollen, as the the result either of transudations or extravasations. Some have thought that all cases of this kind were due to injury, to falls, or to blows. The truth probably is that neither cases of hæmatoma auris, or the frequently-occurring fractured ribs among the insane, are always or often due to cruel treatment; but that, in handling such patients without due care or consideration, slight injuries produce these affections to which the patients are predisposed, and which might have occurred spontaneously if the injuries had not been inflicted.

It is sometimes said that attendants should be encouraged to take a scientific interest in their patients. Properly understood, this assertion is true. I do not think that it will be of any service to the unfortunate patient for the attendant to become a sciolist in psychiatry,—that is, to have a smattering of knowledge of anatomy, physiology, and pathology of insanity to parade on occasions or make use of improperly in behalf of the patient; but the kind of scientific knowledge which attendants should have in order to manage the insane for their best interests is that which comes from a study of the mental, physical, and other qualities of their patients, and of the details of the best methods of handling them mentally, physically, and otherwise. They should study their patients; they should learn what gives them pleasure; what, in cases of melancholia, will best serve to draw them out of the mental shell into which they have retired; in cases of mania what will best answer to soothe and calm their stormy minds; what, in cases of fixed and limited delusions, will best divert them from the unhealthy channels into which their thoughts are directed.

Nurses and attendants should have sufficient knowledge of what is meant by delusions, hallucinations, illusions, etc., to enable them to understand the mental condition of the patient, for certain practical purposes which come within their own province. Not a few insane have delusions which they, for a time or altogether, suppress, or which are not freely or openly expressed. The existence of such delusions in most cases is, indeed, must be, known to the physician who has the patient in charge, and the nurse or attendant should be promptly informed, in order not only to guard against special dangers which may arise out of the delusions, but also to be better able to manage the unfortunate patient without offence or injury. Thus, an attendant who knows that the patient has delusions which may lead to mutilation, will be on guard against the occurrence of mutilation. In one case under my care the patient's delusions sometimes took the form that it was his duty to destroy his eyes, because they had offended, and he almost succeeded in accomplishing this insane purpose before a new attendant had become properly informed as to the delusion.

The possibility of the patient mutilating himself or herself should always be borne in mind. Under the influence of peculiar delusions mutilations of a most extraordinary character are sometimes performed, and attendants should always be upon the alert. To no one as much as to the insane does the old saying, that it is the unexpected that always happens, so fully apply. After seemingly almost every avenue for the commission of suicide or self-mutilation has been provided against, some new device will be put in action, and perhaps successfully. I recall one instance of a patient in a large hospital who was noticed speaking in a mumbling manner. He was at first passed by, but, on second thought, was questioned, and on refusing to open his mouth it was opened forcibly, and it was found that he had tied his tongue with a cord, which had caused it to be much swollen, and would probably in a short time have brought about sloughing and possibly fatal hemorrhage. In this case the act was performed in all probability under the influence of a delusion that his tongue was an unruly member. Although not under my charge at the time of the occurrence, this man had been a patient of mine, and one of his delusions was with reference to his tongue or speech.

It is well that attendants should know not only that the sane sometimes simulate insanity, but that the *insane* also sometimes simulate insanity, giving a curious combination of a real and a feigned disease which is sufficient sometimes to baffle a skilful expert. It is well that nurses and attendants should appreciate this, because it has happened to me, as well as to other physicians who are concerned with the care and treatment of the insane, that attendants, otherwise faithful and suitable for the work, become disturbed as to the existence of insanity in the patients whom they discover to be shamming in some particulars. They discover some deception, and with their little knowledge begin to think that they have solved the problem of the case better than the doctor himself.

Still another similar point, which it is well, perhaps, that attendants should know, is that among the really insane hysteria in some form is often present. We have not only a form of insanity which is called hysterical, but hysteria in many forms manifests itself among the insane of different classes. I am led to remarks like these for practical reasons which have grown out of experience. I have had a faithful nurse come to me and say that a patient was hysterical, and only wished to create a sensation; that she did not believe that the patient would hurt herself; and such statements may have had some foundation in truth : but in just such a case as this, also, I have known, within twenty-four hours of the time at which the statement was made, the patient to almost succeed in committing suicide in a way that left no doubt for a moment even in the mind of the skeptical nurse. A nurse should not have an opinion which she puts in practice to the possible danger or detriment of patient or physician. In the long run, she will be safest to refer all doubts to the physician, rather than to act upon them herself.

Later I will speak of the importance of cleanliness on the part of attendants, and also of the close attention which they should pay to cleanliness on the part of patients. Not only should cleanliness of body be attended to as a matter of great importance, but cleanness of speech and of heart are essentials for nurses and attendants upon the insane. A vulgar, profane, or immoral person has no right to be in charge of an insane patient. Opportunities for evil and for injury are great, and should be guarded against by carefully selecting attendants of high moral tone.

If I were to be asked who make the best attendants, I should say young men or women, not necessarily overlearned in medicine, but bright, intelligent, sympathetic, and sensible persons. At one of the large asylums, in answer to my inquiries, I was told that on the women's side young, healthy Irish girls, on the whole, made their best attendants, for the reason that, as a rule, they were strong, had good animal spirits, were quick, and at the same time were kind-hearted.

One defect serious in character, however, was a tendency not always to adhere to the truth in their statements. A good nurse or attendant for the insane should be firm and yet forbearing, should be quick and yet not hasty, should be courageous and yet not aggressive, should have tact and yet not be deceitful. It has often been observed in large institutions for the insane that the coming of a new nurse or set of nurses will change for a time the conduct and character of a whole ward. Order will give place to confusion and quietness to discord. A nurse who is fussy and too demonstrative is, above all, a nuisance, particularly in public institutions. Nurses should avoid lying to or deceiving the patients under their charge. Just as in the examination of the insane an occasion may arise when a doctor will be justified in resorting to stratagem or even to deception, so it is possible that a nurse or attendant in some grave emergency or peculiar position may be justified in temporarily deceiving the patient. Such occasions are, however, exceedingly rare, and deception should never be practised when it is in any way possible to avoid it. Most insane patients, unless they be dements, are capable of recognizing deceit and dissimulation in others. The tendency of the insane mind itself is often to deceit and dissimulation, and a part of the moral or psychical treatment of the insane should be the enforcement by example of physicians and nurses of good moral principle. Promises which cannot be kept should not be made.

The duties of a faithful nurse or attendant upon the insane are, as a rule, arduous and wearing. Sometimes a nurse may be put in charge of a private case, one of dementia or mild melancholia for instance, which will give comparatively little trouble; but, on the whole, cases of insanity require great care and attention, and to be in charge of them is not to hold a sinecure. In one of the large State hospitals, with which I was formerly connected as consultant, the best and strongest attendants, after a year of faithful service, usually showed marked signs of physical and mental wear. The amount of service required by the regulations of some of the large asylums is, I think, too great, while, as a rule, the compensation is far too small.

The nursing and care of the insane in private practice is a matter of grave importance and one which requires some special points of consideration. If sufficient means are at hand, some forms of insanity can be as well or even better treated at their own homes or in private houses than in hospitals or asylums, either private or public. Some forms of mental disorder are comparatively easy to manage at their own homes. Most of them, however, present peculiar difficulties, which call for particular mention and attention. Among the cases most likely to be treated with success at their homes with ultimate chances of recovery are mild melancholia, common acute mania, when not too severe in character, and some forms of hysterical insanity. Some other forms of insanity, such as senile and secondary dementia, idiocy, and imbecility of various grades, may be cared for at home, as a rule, without danger, although in such cases nothing is to be hoped for so far as cure is concerned. The effort should never be made to treat patients at home or in private houses if such patients require constant watching, unless the friends or relatives of the patient are fully able to employ faithful attendants. The physician is often asked whether such and such members of a family cannot, as well as not, look after the patient. As a rule, but not one without exceptions, it is running considerable risk to allow a patient who is dangerous to himself or others to be attended solely by members of his family. It certainly cannot be done without great risk if those who are supposed to have charge of the patient have other work or duties which claim their attention. When the patients are so situated financially that attendants cannot be hired, or that members of the family or friends cannot give them exclusive attention, it is far better to resort to hospital or asylum care.

It will frequently happen in the private or home care of the insane that two attendants are put in charge of a patient; occasionally the number will even be greater. In all such cases, unless the attendance is simply such as requires one attendant to be in charge absolutely for a certain number of hours and the other for a like time, the physician finds it desirable to have one in chief authority. To one nurse he consigns the chief charge, making him especially responsible, and requiring the other, when it is necessary, to act under instructions. Occasions frequently arise, particularly in the care of the violent or suicidal insane, when it is necessary for prompt and decided action,—such action as can be taken and directed by one clear head. In such a case it should be understood that the requests or even directions of the chief nurse shall be coincided with or obeyed by the other. I have known a patient to be injured, so far as efforts to control him were concerned, by the petty jealousies which have arisen between the nurses in attendance. It has seemed to me —possibly I may be mistaken with reference to this— that these are more likely to arise among women than among men. No appearance of discord or disagreement between attendants should ever be shown before a patient. Such an exhibition is in some cases destructive of all efforts to control a patient for his or her own good.

Much of the instruction which is given to those who are in charge of the insane, whether resident physician, nurses, or attendants, must necessarily have reference to the habits, conduct, and peculiarity both of patients and those in charge; and it is for this reason that I am compelled to devote so much space to the consideration of questions of this kind.

Attendants should never foolishly ridicule patients. While sometimes harmless amusement may be derived from listening to the vagaries of the insane whose delusions or notions are of a pleasant or joyous type, on the other hand patients should never either be encouraged in their delusions, or ridiculed because of them or of anything else which is the result of their impaired mental condition.

In institutions, nurses and attendants should be careful not to indulge in favoritism, and not to single out any patients for special antipathy. Probably nowhere in the world can so many annoying, aggravating, and generally disagreeable individuals be found together as in a large insane hospital. It follows as a matter of course from their mental condition that this should be so; but the highest type of nurse or attendant is one who, appreciating all this, is able to pursue a straightforward, honorable course in caring for every patient.

All disputes and differences of opinion should be referred to the physician. The highest tests of the quality of a good attendant in cases of mental disorder are those tests which are exhibited in the personal characteristics of the individual. Forbearance under temptation, the ability to suppress envy, jealousy, or self in any way, the willingness to do all things and deny all things for the sake of the patient, furnish the best evidences of the needed qualities. The nurse who is given chief authority in the care of a private patient should not be officious or exacting, but should make it her special aim to see that the directions of the physician are implicitly carried out; should report to the physician in charge everything disagreeable, and all derelictions.

In most cases it is well for the chief nurse to keep a careful record of the details of treatment and also of the meals, amount of sleep, exercise, kind of occupation or amusement, etc. A systematic method of recording the time of administration of medicines, and so on, is advisable. Printed blanks are used in some institutions, and if carefully prepared may be of great service.

Attendants upon the insane should not become to any great extent routinists. They should not become so familiar with their patients as to forget many of the little things that they may need. The insane often suffer for want of little things,—for things of which they need not be deprived and of which no one cares to deprive them, but which they often do not get because of carelessness or thoughtlessness. They often suffer, for example, for water. I have been told by an intelligent lady that she drank her own urine, not because of a delusion, but because of absolute uncontrollable thirst which was not properly gratified. Such probably is the case in not infrequent instances, although acts of this kind are more frequently, no doubt, due to delusions.

Sometimes the insane do not receive sufficient food. In some forms of insanity, owing to the tremendous waste of tissue which takes place in consequence of the wearing mental disorder, large quantities of food are called for. Of course, arranging the amount of food is a matter for the physician, but nurses should see to it that the food is given as directed. As a rule, insane patients eat too much rather than too little, and this, of course, must also be guarded against.

The protection of the insane from the commission of suicide is a most important duty upon the part of an attendant. It is strange in how many ways and with how much shrewdness the insane will accomplish this dire purpose if intelligent vigilance is not exercised. Suicides by the insane, like suicides by the sane, are of the most diverse character, as far as method of performance is concerned. The attempt to hang is very common. An insane patient has been known to hang herself by a shoe-string or an apron-string to the doorknob of a room, or the bed-cords, the sheets, clothing, etc., have been employed. To drown themselves, patients do not require rivers nor even bath-tubs. They have been known successfully to accomplish this purpose by thrusting the head into a pail or even into a basin of water. Sometimes objects are thrust by the patients into their own mouths to choke themselves or strangle themselves to death. Sometimes they cut themselves with glass which has obtained by breaking windows.

Knives, scissors, razors, and all forms of cutting instruments should be carefully kept out of the way of patients who are likely to injure themselves by such means. The greatest possible care should be taken by nurses or attendants with reference to the custody of medicines administered to the patients. Here, again, it is advisable that they should have some little practical instruction as to the strength and danger of the most potent remedies which are intrusted to their care for administration. They cannot be expected to have exact and profound knowledge as to the physiological properties, sources, etc., of morphia, conium, hydrobromate of hyoscine, atropia, duboisia, digitalis, chloral, bromides, cannabis indica, and the long list of powerful medical implements which are put into their hands; but they should know that these drugs are dangerous and deadly poison in more than a proper dose. They should keep them out of the reach of patients in intervals between the times of giving the particular doses ordered; where possible, they should be kept in an adjoining room, and always under lock and key and entirely out of reach. While, however, in the case of patients suicidally inclined the means of suicide should be kept or put out of reach, this should not be done in a fussy or demonstrative way. Such things should, so far as possible, not attract the attention of patients.

Comparatively little restraint of a mechanical kind is now used in the care of the insane. In some cases, however, rare though they be, it may be absolutely necessary to protect patients from mutilation or suicide. Mechanical restraint should not be applied by a nurse or attendant except by the direction of the physician; and, if it is used, the greatest possible care should be taken that no evil consequences result. Not long since, in a large hospital for the insane, a patient who had been restrained by straps in a bed, and was supposed by the attendant to be sleeping, so dragged himself down over the side of the bed as to actually cause his death by means of the straps with which he was supposed to be restrained and protected.

When it becomes absolutely necessary for a nurse or attendant to take hold of an insane patient, care should be exercised as to the manner in which this is done. Usually patients requiring to be taken hold of are violent,—cases of mania, homicidal monomania, of agitated melancholia, and the like. When two attendants are present it is well, while one secures the attention of the patient by demonstrations in front, for the other to get behind him and either throw the arms around him or seize both arms carefully but firmly above the elbows. Whatever plan of seizing patients is adopted, great care should be taken not to inflict injury. It is only in self-defence, that is, in defence of life or limb in a great emergency, that an attendant is ever justified in striking or roughly handling a patient. A patient, of course, may kick backwards when seized from behind; but the attendant by slipping his feet and legs as far as possible to one side can usually avoid being injured in this way.

When applications of heat or cold are made to the insane, special attention should be paid to them. In some forms of insanity great insensibility of the skin is present, and patients would in these cases allow blisters or mustard-plasters or stupes, or fomentations or any form of hot or counter-irritating applications to remain until serious blistering occurred. On the other hand, the insane are sometimes more impatient of such applications than the sane; in any case, special attention should be given to the matter by the attendant. The same remarks apply to the use of cold water or ice.

All physicians who are engaged in the care and treatment of the insane now recognize the importance of occupation, amusement, and educational training as therapeutic resources. To attendants, the carrying out of plans of occupation, amusement, and teaching must be largely intrusted. In institutions, of course, methods and measures for the promotion of these objects will be carefully planned and regulated, if the institutions are up to the times in their management. The great curse of many institutions for the insane has been idleness. It is not within my design in the present chapter to go into a consideration of the methods and forms of employment, amusement, and instruction, but simply to give a few hints and suggestions as to the manner of carrying out the details, so far as such details are intrusted to nurses and attendants.

In the first place, patients should be encouraged, and in some cases be compelled, to do that work which is called for in their own personal care. They can be encouraged, or made, to take care of themselves, so far as cleanliness is concerned. Female patients should be made, when possible, to care for their own rooms, beds, etc. Attention should be paid, of course, particularly in private practice, to the question of the usual customs and habits of the patient.

No matter what may be the work at which an insane patient is engaged, whether sewing or knitting or quilting or ironing or book-keeping; or, if men, gardening, ploughing, brush-making, chair-making, printing, or type-writing, a duty that will be incumbent upon the attendant more often than upon the physician will be to see that the patient does not overwork himself. While the tendency will be to do too little, in some cases insane patients will certainly try to do too much. This is particularly true in some forms of insanity attended with excitement, some varieties of mania, for instance, in which the patient tends to do everything to excess. Again, a patient may overwork as the result of some special delusion. Thus, Dr. Massey[1] speaks of the case of a lady in good circumstances who, having been left with a family to support, resorted to dress-making, but became insane and was sent to an asylum, where she was induced to go to work in the sewing-room. The result

---

[1] *Penn Monthly*, November, 1879.

was the production of a new delusion: she believed she was in charge of the sewing-room with large wages, and made great exertions in consequence. Becoming troublesome, she was transferred to the laundry, where the same thing was re-enacted; this resulting at last in complete exhaustion and confinement to bed.

Exercise is of the utmost importance to many insane patients. In suitable weather, out-of-door exercise should be carefully provided for; and here, again, the careful attendant will guard against under-exercising on the one hand, or over-exercising on the other.

Torpor of the bowels and obstinate constipation are of such frequent occurrence among the insane that both resident physicians and attendants should have their attention particularly directed to these conditions. In cases of melancholia, the bowels may remain constipated for days and even weeks if the patients are neglected. A large number of insane patients in asylums, particularly many of the cases of dementia, are troubled with involuntary evacuations from the bowels or bladder, particularly at night. In private patients the same difficulty may be encountered, and it is well for physicians and nurses to have some plan of meeting the difficulty. In one large hospital the plan adopted is a very sensible one. The physician in charge has a list of all patients who are troubled in this way, and in the evenings before retiring injections of warm water are given; the lower bowel is thus emptied, and the probability is that the patients will not soil their beds. In some cases, in addition to emptying the bowels, the bladder may have to be relieved by the catheter.

Cleanliness on the part of a nurse or attendant upon the insane is of the utmost importance. It is important not only for itself, but for the example which it sets to the patient. The insane may be made more filthy than their mental affliction would lead them to be by the example set to them by others; on the other hand, they may become cleanly in habit simply from an example steadily set.

The bathing of insane patients is a matter of great importance, both from the point of view of cleanliness and as a therapeutic agent, and attendants should be thoroughly instructed with reference to this matter. They should also have a lively appreciation of the dangers which may attend bathing. A terrible accident recently occurred in one of the larger asylums of a neighboring State, the patient having been literally parboiled by an attendant. As many insane patients make a great fuss and offer much resistance to bathing at all times, a careless or impatient attendant might fail to listen or assure himself of the conditions. Rigid rules are prescribed in all well-regulated public institutions. The rules for bathing which are in force at the State Hospital for the Insane, Norristown, Pennsylvania, so fully express what should be said with reference to this matter that I will quote the most important of them:

> The ward captains will personally supervise the bathing of patients, which shall not be conducted during their absence without permission.
>
> Every patient is to be bathed immediately after admission, and once a week afterwards, unless excused by medical order. Should

there be the least doubt as to the advisability of bathing any patient, owing to sickness, feebleness, or excitement, the matter should be immediately reported to the medical officer.

To provide against catching cold, the captains will see that the bath- and dressing-rooms are sufficiently warmed at bathing times, otherwise to postpone bathing until the rooms are heated.

Any marks, bruises, wounds, sores, pain, or evidence of disease complained of by the patient, or noticed during any of the bathing operations, must be immediately reported to the physician.

During the use of the bath, the room is never to be left by the attendant, except by special permission of the resident physician. When the room is not in use the door must be kept locked.

Before putting the patient into the bath observe that the water is of proper temperature. It should not be less than 88° nor above 98°.

Never turn on the hot water when the patient is in the tub.

In the bath the body of the patient is to be well cleansed with soap; and in washing the hair be careful that no soap gets into the patient's eyes. After leaving the water, especial care must be taken to thoroughly dry the patients, and clothe them as rapidly as possible. A separate towel must be provided for every one.

Under no pretence whatever is a patient's head to be put under water.

An attendant must not attempt under any circumstances to bathe a struggling patient alone.

Cold baths must never be given.

Neither before nor after the bath will patients be allowed to stand about unclothed.

The keys are never to be left on the bath-taps, nor are they to be used by the patients.

I will next speak briefly of the methods of feeding patients forcibly. I do this because this book is intended for the instruction of resident physicians, as well as of nurses and attendants. I do not believe that nurses or attendants should be allowed to have the feeding of patients in this manner under their own charge; but, as it is necessary for them to assist in performing this work, they should receive instruction as to the whole process in order that they may be better able to render assistance to the physician. In some institutions head nurses have been allowed to administer food in this way; but it is contrary to the regulations of all the institutions for the insane of which I have knowledge to allow this to be done.

Not a few cases of insanity require, at times, to be fed by force. The cases in which such treatment is called for differ somewhat. In melancholia patients will allow themselves to rapidly run down in health, or even starve to death, if the food is not promptly administered by artificial means. Sometimes cases of mania will refuse to take food, either because of their general excitement and combativeness or because of delusions. Paranoiacs or monomaniacs refuse food, if at all, because of some fixed delusion. A case of chronic alcoholic insanity, or even in rare instances a paretic dement, may refuse food because of a delusion, as, for instance, that food is being poisoned by physicians, attendants, or others, or that it is a means resorted to to accomplish some purpose on the part of a persecutor. The refusal of food by the case of hysterical insanity, like the apparent attempts at suicide, is usually simply resorted to for the purpose of exciting sympathy. In most cases these patients will obtain food on the sly if possible, but sometimes they will carry on the deception sufficiently long to injure themselves, if the attendants do not resort to active interference.

Various plans of feeding patients by force have been

recommended at different times; two of these are now almost universally employed, and one of the two, namely, nasal feeding, is now coming more and more into general use.

The three methods in more or less common use will be described. The first plan is that of holding the nostrils closed and opening the mouth, or thus compelling it to be opened, while the food is quickly poured into the mouth, and the patient is forced to instantaneously swallow it. A certain amount of danger accompanies this plan, which, by the way, is the one often resorted to in the case of recalcitrant children. If any one should try the experiment of trying to swallow liquids or solids while the nose is tightly held, so that no air can enter through it, he will find that it is a very difficult or almost impossible thing to accomplish. The very fact that the reason that a patient or individual, whose nose is held, opens the mouth in order to breathe shows the source of possible danger, which is that the glottis being open in order to receive air into the lungs at the same moment that food is hastily thrust into the mouth, some of the food is very likely to find its way into the windpipe and thus choke or strangle, or partially choke or strangle, the patient. This method of feeding is now not much employed.

Dr. D. Anderson Moxey[1] was among the first to call attention to the administration of food and medicine by the nose when they could not be given by the mouth. He resorted to this method in cases of insanity, in in-

---

[1] The *Lancet*, March 20, 1889.

flamed and ulcerated sore throat, in glossitis, in deep intoxication, and in infancy, where nothing could be administered by the mouth or retained by the rectum. In cases of insanity he fed his patients without the use of a tube by simply holding a small wedgewood funnel in one of the nostrils, and pouring through it the liquid or semi-solid nourishment. He describes his method with insane patients as follows: After first trying to induce the patient to take the nourishment quietly and in the ordinary way, he summoned three attendants at least (four or five, if they could be had), and laid the patient down on his back as quietly as possible on a low couch in the middle of the room. If there were only three attendants, one controlled the legs and the other two the arms; if a fourth was present, he attended only to the head, which was held between his knees as the attendant sat on a low stool at the top of the couch. If a fifth was present, he assisted in holding down the legs. In this way the patient was completely controlled, and sometimes would yield and swallow in the usual way. Then introducing the end of the funnel gently into one of the nostrils, he poured the liquid slowly into it, pausing now and then to allow the patient to take a deep inspiration, and not allowing the food to accumulate in the funnel. A determined patient would generally be able to sputter a little of it out of his mouth; when such was the case he poured the contents of the jug faster into the funnel, and sometimes obstructed the other nostril. In troublesome cases a medical man ought, he insisted, to invariably administer the draught, as he alone could properly judge of the extent to which it was necessary to interrupt nasal respiration. Dr. Moxey never found any serious results from this method of feeding. I have referred to these observations of Dr. Moxey to show the perfect feasibility of nasal feeding, and one of the methods of holding the patient by which it can even be accomplished without the aid of a tube, although undoubtedly the use of the tube is the neater and better method.

A good tube for feeding by the nose can be made by attaching a soft Nélaton catheter to a long piece of drainage-tube of a calibre which will just allow the catheter to tightly fit into it. To the extremity of the drainage-tube a funnel of large size should be attached.

In nasal feeding some of the difficulties and dangers which are met with should be borne in mind. The number of patients who cannot be fed by the nose is very small; occasionally, however, a patient is found whom it seems impossible to feed in this way, owing to the choking and strangling produced. This may be because of some peculiar anatomical conformation, or some special idiosyncrasy on the part of the patient. Such a patient will choke or strangle with nasal feeding when he will not when the stomach-tube is resorted to. If, when the attempt is made to pass the well-oiled tube through the nostril, resistance is encountered, and if, after a few trials, the tube cannot be made to pass, great force should not be employed by the operator, but the tube should be at once withdrawn and the effort should be made to pass it through the other nostril. In nearly all cases where special resistance is offered on one side the tube will pass with ease upon the other, and this, in most instances, is because, if hypertrophies or projections exist upon one side, there will be upon the other corresponding or compensating depressions and enlargements. Sometimes, but rarely, the mucous membrane is exceedingly irritable. After the nasal tube has passed through the nostrils it seems to have the peculiar tendency in some cases to drop into the glottis, the patient struggling and attempting to scream meanwhile. Some patients will spit or force the tube out into the mouth, and attendants can sometimes, through the mouth, keep the tube, which has been passed through the nose, in position. Occasionally the nose is made sore by the use of the tube, but this is not likely to occur if the tube is always perfectly cleaned and well oiled. If it is of the proper kind,— that is, a soft tube,—there will be no danger of injuring the parts by breaking or perforating the mucous membrane. Indeed, one of the advantages of the nasal tube over the form of stomach-tube or œsophageal tube that is commonly employed is that the danger of injury by perforation or abrasion is much less. The stomach-tube must be a little larger and of firmer make, and it is likely, after a little usage, particularly if not looked after with great care, to become stiff and hard. I know of one instance in which the œsophagus was pierced by a stomach-tube in the hands of an unskilful attendant. In using the nasal tube, great care should always be exercised to see that at least fifteen to sixteen inches of the tube has been passed before beginning the feeding. This will make it certain that the entrance to the windpipe has been passed. Of course, care should be taken to observe that the tube has not doubled on itself. Dr. Spitzka speaks of the method of feeding by force as follows:

"Whether fed with the funnel or the stomach-pump, the patient should sit up; and if he is very obstructive, a restraining-chair will save the patient much needless muscular exertion, the physician much trouble, and diminish the chances of doing an injury. In case the œsophageal tube is passed along the floor of the nasal cavity, it is apt to encounter a resistance and be deflected forward by a prominence which is sometimes very marked on the posterior pharyngeal wall, and which corresponds to the bodies of the cervical vertebræ. Dr. Tuke advises throwing the head of the patient back at the moment when the sound approaches the posterior nares, the tube having previously been bent a little so as to facilitate its downward passage; then at the moment when it is about to glide down into the œsophagus, when there is a risk of its passing into the larynx, he advises the head to be brought forward and downward so as to send the point against the posterior wall of the pharynx. After passing the upper end of the œsophagus, the tube is usually swallowed, as it were, and glides down without any further difficulty into the stomach through the action of the constrictor muscles."[1]

Whenever a patient is fed forcibly, care should be taken to have ample force to restrain him without

---

[1] "Insanity, Its Classification, Diagnosis, and Treatment." By E. C. Spitzka, M.D.

a violent struggle. The sight of overwhelming force will have a good moral or psychical influence over the patient. Three persons can feed a patient successfully, no matter how great the inclination to resist, if these persons are skilful and have sufficient strength. With less than three persons the difficulty will be very great, and in some cases it will be impossible successfully to accomplish the purpose. One attendant should always be charged with the task of firmly holding the head in position; another can hold the arms and hands, with the body thrown over the limbs of the patient, who before this is attempted should be well enveloped in a blanket or sheet. It is much better, however, to have two attendants hold the arms and legs of the patient.

After a patient has been fed, the tube should be always promptly cleansed. An attendant who does not attend to this matter is derelict in duty. Hot water cannot be used because of the melting of the india-rubber, but the tube should be washed by allowing cold water to run through and over it. Negligent attendants are likely to neglect this matter.

Whether fed by the mouth and œsophagus or through the nose, patients in rare instances learn to vomit or regurgitate the food. If this occurs, it is a matter for a physician rather than for a nurse or attendant. It has been found that the administration of morphia and hyoscyamine prior to the time of feeding will sometimes prevent the patient from exercising this power. Two doses, say of sulphate of morphia gr. $\frac{1}{4}$ and hyoscyamine gr. $\frac{1}{60}$ to $\frac{1}{80}$, may be given, the first three or four hours, and the second one hour or less before the times for administering the food.

If the nasal mucous membrane or the mucous membrane of the throat should be unusually irritable, resort may be had to a weak solution of the hydrochlorate of cocaine, which can be painted over them.

The administration of such remedies as morphia, hyoscyamine, and the bromides will also, of course, in cases of insanity, as in other cases, tend to diminish the pharyngeal reflex and thus allow of the use of forcible mechanical feeding with less trouble; but prescribing medicines is the duty of the physician, and therefore little is said here about such matters.

One advantage in using the stomach-tube through the mouth is that when the latter is kept well opened by means of a bandage or a screw-wedge, the finger of the physician or attendant can be used to guide the tube past the epiglottis into the œsophagus, and thus prevent any danger no matter how much the patient may struggle or use his breathing apparatus.

No matter how a patient is fed, whether by the nose or through the mouth, care should be taken to have everything ready before the operation begins. It is one of those processes which cannot be done properly if it is only half prepared for or attended to indifferently. The proper appliances should be on hand and within easy reach. The food and medicines that may become necessary, if they have been previously ordered, should be ready for the doctor. Great care should be taken in the preparation of the food, which is administered artificially, if it is anything more than milk or some single substance of a fluid kind. If gruel or any form of semi-solid food, care should be taken to have it of the exact consistence, which is best administered through a tube. When medicines are ordered to be given with the food, as for instance bromide of potassium in milk, only just as much as is required should be mixed at one time.

Even when patients are not fed by force, it is of importance that their eating should be carefully watched and attended to by those in charge. In some forms of insanity the tendency is to bolt the food with great rapidity and in enormous quantity. So little control have some demented patients over matters of this kind that they will choke and strangle themselves, or fill themselves to more than repletion. Such patients should, by kind but firm measures, be trained to eat more slowly and carefully.

It is sometimes desirable that the temperature of the insane should be taken, and when this is the case special precautions should be used. It would, for instance, ordinarily, be foolish to attempt to take the temperature of an insane patient in the mouth. In many cases the thermometer would either be broken in the struggle which would take place, or it might be bitten off by the patient, and the glass swallowed with fatal results. Only in cases where the exact mental status of the patient is so well known that the physician, feeling that no danger can arise by taking the temperature by the mouth, directs it to be done, should this plan be followed. The temperature of the insane should therefore usually be taken in the axilla, and when taken in this way the instrument should not be left so that the patient can take hold of it himself. The nurse should stay by the side of the patient, and by preference in most cases should hold the thermometer in position the whole time of the determination. Occasionally, when specially directed so to do, the temperature may be taken in the rectum, but here, as in the mouth, great caution is usually necessary, and it is often advisable not to resort to this method.

Some valuable investigations as to the temperature of the head in the different forms of insanity have been made, and nurses and attendants should be informed as to special requirements for this procedure. (See pages 60-62.)

The following are some of the most important rules enforced in the Female and in the Male Departments of the State Hospital for the Insane, Norristown, Pennsylvania:

Attendants are expected to devote their whole time to the performance of their accepted duties, in which they will be under the immediate and constant supervision of the head nurses; they will not leave their wards except by permission, and under no circumstances shall a ward be left without an attendant. They will not visit another ward without permission.

Attendants will never loan their keys, nor allow any unauthorized person to enter their wards without permission of the physicians; they will avoid talking to visitors and others about the patients, their names, peculiarities, etc., referring all inquiries of the kind to the physicians; they will not visit or correspond with the friends of patients, or mail letters for patients, without permission from the physicians.

Each attendant is responsible at all times for every patient under her charge, and in the event of an escape traceable to want of care on the part of the attendant, the latter will be held responsible for the expenses of recovery. Every injury to a patient, however, received, and any unusual symptoms, shall be reported to the head nurse or medical officer.

No attendant will be excused, under any circumstances, for striking, choking, kicking, or otherwise maltreating a patient, and no one will be excused for failing to report promptly any such case that comes to her knowledge. Physical force is to be made use of only when unavoidable, and is to be looked upon as an unfortunate necessity, to be at once reported to the head nurse, or other officer. Attendants are not to ridicule, scold, threaten, or speak rudely to patients, but are to strive to win their respect and affection by continued kindness and forbearance.

No attendant will be excused for receiving money from the friends of patients; they will not accept gifts of any kind from patients or their friends without permission of the physicians; they will not wear, or make use of, any clothing or property belonging to a patient.

Attendants will not seclude patients except under direction of the head nurse or other officer; if necessary, in an exceptional case, to separate a patient from others at once, the facts shall be reported without delay.

Attendants are responsible for the safekeeping of all property in their respective wards. If anything be unavoidably destroyed, they shall report the facts, which will be recorded in the "Damage-Book"; if destroyed or lost by their own want of care, they may be held responsible for costs; they are enjoined to exercise special care in the use of scissors, knives, and all instruments of possible danger, and in case such an article is missing from the known number, to report the fact without delay.

Attendants may be excused from duty by the head nurse under direction of the supervisor or physicians. In going out, attendants will leave their keys at the central office; both in going out and coming in they will invariably report to the head nurse. In case of resignation, two weeks' notice is required; the same notice will be given in case an attendant is considered unsuited to her work, but a flagrant violation of rules or breach of duty may be followed by immediate discharge.

Attendants must set a good example, never scold, threaten nor dictate, but always be gentle and patient. They will never allow patients to be laughed at, ridiculed, nor harshly spoken to on account of their delusions or the peculiarities of their conduct.

All cases of abuse and maltreatment are required to be reported to the Lunacy Committee, and attendants thus reported are liable to be prosecuted according to law.

Attendants will not retire to their rooms while the patients are in the wards, nor allow patients to go into attendants' rooms.

Neatness and cleanliness will be most scrupulously regarded by attendants, not only in regard to their own appearance but in the patients and wards under their charge, and any failure in these will be considered a neglect of duty.

No patient shall be taken out of the wards for any purpose, or by any person, without the knowledge of the ward captain or supervisor, and when taken out, he must be returned by the same person.

At meals one or more attendants will always be present to wait upon the patients.

Knives and forks must be counted after each meal, and the attendants must see that they are not carried away from the table.

Attendants shall in no case visit friends of patients, nor receive from them money or presents without the knowledge of the physicians.

A uniform temperature of seventy degrees in the halls will be required in the cold season.

The attendants will regulate the temperature by opening and closing the heat registers in the halls, and those in the bedrooms, which are located between the upper and lower ventilating registers.

The wards may be ventilated in the early morning between the hours of five and six, by raising the windows, but throughout the remainder of the day care must be taken to keep them closed.

The captains will see that the doors in the passage-ways leading to the basement are kept closed during the day.

The captains must instruct the attendants in the control of the artificial ventilation of the wards, which is accomplished by means of the two sets of ventilating flues.

The ventilating register near the floor of the rooms must always be kept open and unobstructed. The ventilating register near the ceiling of the rooms must be opened only when the air is very foul, or to reduce the temperature.

In the spring, when the artificial heat is taken off, the upper registers are to be kept constantly open.

The captains must also instruct the attendants in the use of the fire-hose.

I will conclude this chapter by quoting in full the quaint description of the "Good Attendant," by Dr. Ray,[1] the greatest of American alienists:

"The good attendant never shirketh his appointed work, and it is not in him to be satisfied with just that measure of performance which will enable him to keep his place. He elevateth his employment by the manner in which he performeth its duties. Though offensive to the senses, or trying to the temper, or exhaustive of patience, as many of them are, yet he meeteth them all faithfully and promptly. Like every true man and true woman, he findeth that dignity inherent in every good work, that ennobles even the meanest service. As the good artisan rejoiceth over some choice specimen of his craft, wrought by his own hand, so doth the good attendant rejoice when, after much toil and trial, he seeth the mind of his patient coming out from under the cloud. To hasten this blessed consummation, he spareth neither time nor trouble, rendering every attention needful for the bodily comfort, and by unceasing arts of kindness soothing the troubled spirit. The good at-

---

[1] "Ideal Characters of the Officers of a Hospital for the Insane." By I. Ray, M.D.

tendant is ever gentle in his words and ways, and under no provocation will he return a blow or an abusive word. Unlike the people of former times, who believed that the insane must first be made to feel that they have a master in their keeper, and for this purpose resorted to threats and blows, he seeketh to obtain the desirable control by gaining the patient's respect, and this he well knoweth will not follow angry words, or harsh measures, or any form of intimidation.

"The good attendant never attempteth to reason his patient out of his false beliefs, and, as far as practicable, he preventeth him from conversing about them. He knoweth that argument giveth them additional strength, besides exciting and souring the temper. He refraineth from joking on the notions or circumstances of his patient, for he hath learned that the disordered mind is impervious to a joke, but rather construes one into an insult. He is careful to obey every change, bodily or mental, for better or worse, and maketh due report thereof to the physicians. His constant presence with the patients giveth him opportunity to see and hear much that may escape the attention of the officers in their casual visits, and his eyes and ears are ever open for this purpose. Especially doth he endeavor to inspire his charge with confidence in the physicians, always holding them up as his friends and protectors, who will never see him wronged or injured. When abroad, he refraineth from entertaining company with the fancies or conduct of his patients, nor is he swift to pour into itching ears the gossip of the house. The rules made for the government of attendants he faithfully follows, bound thereto by a sense of respect for himself and of fair dealing with his employers.

"The good attendant avoideth all vulgar ways in language, dress, or demeanor, as well as all familiarities which he would never venture upon outside of the hospital. He beareth in mind that the people who have fallen to his charge, however perverted or degraded by disease, were once as good as himself, if not better, and have done nothing to forfeit their claims to his respect and protection. For deficiencies of culture and of good breeding, he more than maketh up by gentle words, acts of kindness, and little attentions. Especially is the female attendant not to add fresh poignancy to the sorrows of her charge by coarse expressions, untidy ways, and manners utterly devoid of refinement."

INDEX.

Air cushions, 33.
Alternate hot and cold applications, 58.
Amalgamating solution, 76.
Amidon's commutator or pole-changer, 84.
Amusement of the insane, 123.
Anderson on tact in nurses, 13.
Apoplexy, general management of, 26.
   hemorrhagic, 26.
   signs and symptoms of, 26.
   what to do in case of, 28.
Applications of heat and cold to the insane, 123.
Attendants, scientific interest of, in the patients, 111.
   tests of good, 119.
   the best, 114.

Barrett's chloride of silver battery, 81.
Bathing, 57.
   dangers of, among the insane, 126.
   of the insane, 126.
   rules for, 126.
Battery, chloride of silver, 81.
Battery fluid, 76.
Beard and Rockwell's method of general faradization, 104.

Bed-sores, 33.
   acute, 34.
   plans of treatment for, 34.
   prevention of, 35.
   treatment with galvanic plates, 35.
Bladder, washing out of, 32.

Cases, different classes of nervous and insane, 11.
   of chronic organic nervous disease, 31.
Cerebral hemorrhage, acute bed-sores in, 34.
Cheyne-Stokes breathing, 27.
Cleanliness of nurses or attendants upon the insane, 126.
Cleansing of feeding-tube, 134.
Combination battery, 80.
Commutator, 84.
   diagram of, 85.
   switch, 86.
Conducting cords, 95.
Conjugate deviation of the eyes and head, 27.
Consciousness, loss of, in epilepsy and in hystero-epilepsy, 21.
Constipation among the insane, 125.
Cotton, use of absorbent, 94.
Current carrier, 92.
   regulators or modifiers, 72.

INDEX.

Defects in nurses and attendants, 115.
Delirium, treatment of, 36.
Delirium tremens, 35.
Delusions, 111.
    suppression of, by patients, 111.
    which lead to mutilation, 112.
Detection of defects in electrical apparatus, 75.
Drunkenness, diagnosis of, 29.
    management of, 26.
    signs and symptoms of, 29.
Dubois-Reymond coil, 68.

Eating, watchfulness of, among the insane, 136.
Educational training of the insane, 123.
Electrical apparatus, importance of keeping clean, 76.
    applications, deep-seated, 96.
    direct and indirect methods of, 97.
    to the brain, 102.
    to the face, 100.
    to the joints, 101.
    to the skin, 100.
    to the spine, 101.
    to timid patients, 74.
Electricity after meals, 99.
    forms of, used in medicine, 64.
    how far it should be used by nurses, 63.
    methods of applying, 96.
    position of limbs and muscles in using, 99.
    what a nurse should know about, 64.
Electrodes for application beneath a patient's clothing, 93.
    method of covering, 94.

Electrodes, 92.
    how to hold, 95.
    methods of cleaning, 93.
Epilepsy, general care of patients suffering from, 24.
Epileptic seizure, aura preceding, 24.
    description of, 22.
    management of, 23.
Epileptics, amount and character of food for, 25.
    peculiar abnormal mental states of, 24.
Eschars, acute, 34.
Estraderé on the varieties of massage, 42.
Exercise for the insane, 125.
Exhaustion produced by electrical applications, 98.
Expedients for amusing and occupying patients, 14.

Fainting, 30.
Faradic apparatus for office table, 70.
    battery, description of, 66.
    Flemming's, 97.
    harm done with, 77.
    hints about handling and caring for, 74.
    current, how to start a, 68.
    machine, other names for, 65.
Faradism, when to use, 99.
Faradization, general, 103.
Favoritism of nurses and attendants, 118.
Food, directions as to its preparation, 36.
    insufficient or too much, among the insane, 120.
    preparation of, for forcible feeding, 135.

INDEX.

Forcible feeding of the insane, 127.
   three methods of, 129.
Frankliuism, 65.

Galvanic batteries, how to put into use, 83.
   battery, care of, 86.
   description of, 78.
   other names for, 65.
   portable, 79.
   current, dangers of, 87.
   injury to sight by, 87.
   regulation of the strength of, 83.
   to head or neck, 87.
Gray's, Dr. Landon Carter, description of the milliampèremetre, 89.
Grease or liniments in massage, 47.

Hæmatoma-auris, 110.
Hallucinations, 111.
Hamilton, Allen McLane, on the use of revulsives, 59.
Heat prostration, 28.
Helplessness, how to be useful in cases of, 32.
Hysteria, 19.
   among the insane, 113.
Hysterical attacks, involuntary, 21.
   simulated, 20.
   cases, general management of, 20.
   seizures, different forms of, 20.
Hystero-epilepsy, 21.
   with separate crises, 22.

Illusions, 111.
Insane, attention to little things among the, 119.
   ear, 110.
   hysteria among the, 113.

Insane, nursing and care of, in private practice, 116.
   patients, methods of seizing or holding, 122.
   ridicule of the, 118.
Insanity, practical hints as to different forms of, 109.
   simulation of insanity by the sane, 113.
   simulation of sanity by the insane, 113.
   what a nurse should know about, 108.
Insensibility, cause and nature of, in various affections, 25.
Interrupting handle electrodes, 93.
Involuntary evacuations among the insane, 125.

Jacoby on the methods of massage, 43, 44, 45.

Kneading, 45.

Lee, Dr. Benjamin, on derivation of word massage, 41.
Levenstein's treatment of narcotic habit, 38.
Loyalty of nurse to the doctor, 15.

Manuals for attendants on the insane, 106.
Massage à friction, 44.
   chapter on, 39.
   complaints of patients as to, 41.
   definitions of, 42.
   derivation of word, 41.
   direction of, 48.
   grease or liniment in, 47.
   immediate, 42.
   mediate, 42.

146 INDEX.

Massage, number of patients to be treated in one day, 49.
   position of muscles and limbs in, 46.
   rough handling of patients in, 50.
   so-called methods of, 42.
   special instruction in, 39.
   temperature and ventilation of room in, 52.
   terms to describe the process of, 46.
   the four essential methods of, 43.
   the use of both hands in, 47.
   twisting or wringing movements in, 48.
   wrist movements in, 47.
Masseur, 46.
Masseurs, mental requirements of, 51.
   knowledge of anatomy and physiology by, 40.
Masseuse, 46.
Meningitis, 36.
Milliampèremetre, 89.
   diagram showing the method of using, 91.
   how to use the, 90.
Mitchell, S. Weir, on a nurse for cases under rest treatment, 14.
   remarks on nurses for nervous patients, 12.
   moral tone of attendants and nurses for the insane, 114.
Motor points, 97.
Movements, 52.
   active, passive, single, and duplicated, 53.
   description of duplicated active, 53.
   design of duplicated active, 55.

Moxey's method of forcible feeding, 129.
Murrell on massage, 41.
Muscle-beaters of Ruebsam, 56.
Muscular applications, hints and cautions about, 98.
Mutilation, delusions of, 112.

Narcotic habit, 37.
Nasal feeding, difficulties and dangers of, 131.
   duties of, 115.
Nurse, decided, 13.
   domineering, 13.
   from a good family, 17.
   the too familiar, 16.
   qualities and qualifications of, for nervous patients, 11.
   the conceited and the too humble, 17.
   the vain, 17.
   who does not talk enough, 16.
   who talks too much, 16.
   who is nursing because she loves the business, 18.
   who makes the diagnosis, 16.
   who quarrels with the servants, 18.
   who will nurse only one type of cases, 19.
Nurses, types of faulty, 16.
   and attendants, defects in, 115.

Observation, habits of, 13.
Occupation of the insane, 123.
Opium habit, 37.
Opium, management of narcotism by, 26.
Overwork by the insane, 124.

Paralysis of the bowels or bladder, 32.

INDEX.

Paralytic cases, 31.
Percussion, 45.
Pétrissage, 45.
Physical condition of masseurs and masseuses, 49.
Playfair on nurse for cases under rest treatment, 15.
Pole, 92.
Pole-changer, 84.

Ray's description of the "Good Attendant," 140.
Recording treatment of insane patients, 119.
Restraint, mechanical, in the care of the insane, 122.
Revulsor, 59.
Rheophore, 92.
Rheotome or current-interrupter of faradic machine, 69.
Rubbing, 44.
Ruebsam, muscle-beaters of, 56.
Rules in force for insane hospitals, 137.

Simulation of insanity, 113.
Sleeplessness, 85.
Southey's, Dr. Reginald, recommendation in supposed drunkenness, 30.
Spencer Wells's treatment of bedsores and ulcers, 35.
Spitzka's description of forcible feeding, 133.
Sponge bath, 57.

Stroking, 43.
Suicide, methods of, among the insane, 120.
  protection of the insane from, 120.
Sunstroke, 28.
Surface thermometers, 59.
  Mattson's, 61.
  methods of using, 60.

Tact, 13.
Tapotement, 45.
Temperature, methods of lowering, 29.
  methods of taking, among the insane, 136.
Tetanus, 36.
Toepler-Holtz electrical machine, 66.

Uræmic coma, 30.
  measures and remedies for treating, 30.
  signs and symptoms of, 30.

Vertigo, caused by galvanic applications, 88.

Water-bed, 34.
Water cushions, 33.
Water-rheostat, 73.
Wet pack, 58.
Work for the insane, 124.

Zinc, directions for amalgamating, 76.

THE END.

SEPTEMBER 1887.

LIST OF NEW AND RECENT WORKS

ISSUED BY

YOUNG J. PENTLAND,

EDINBURGH.

18mo, *Cloth limp, pp.* xii., 120, *Price* 3s.

SYNOPSIS OF THERAPEUTICS,
ARRANGED FOR THE USE OF PRESCRIBERS:

WITH

POSOLOGICAL TABLE AND AN ARRANGEMENT OF THE POISONS.

By R. S. AITCHISON, M.B., Edin.

(1886.)

Large 8vo, Cloth, pp. viii., 325, *Price* 12s. 6d.

CLINICAL STUDIES ON DISEASES OF THE EYE,

INCLUDING THOSE OF THE CONJUNCTIVA, CORNEA, SCLEROTIC, IRIS, AND CILIARY BODY.

By Dr. F. RITTER VON ARLT,
PROFESSOR OF OPHTHALMOLOGY IN VIENNA.

Translated by Dr. LYMAN WARE,
SURGEON TO THE ILLINOIS CHARITABLE EYE AND EAR INFIRMARY; OPHTHALMIC SURGEON TO THE PRESBYTERIAN HOSPITAL, AND TO THE PROTESTANT ORPHAN ASYLUM, CHICAGO.

(1885.)

8vo, Cloth, pp. viii., 374, *with* 408 *Illustrations, finely engraved on Wood,* *Price* 10s. 6d.

TEXT-BOOK OF GENERAL BOTANY.

By Dr. W. J. BEHRENS.

TRANSLATION FROM THE SECOND GERMAN EDITION.

Revised by PATRICK GEDDES, F.R.S.E.,
DEMONSTRATOR OF BOTANY IN THE UNIVERSITY OF EDINBURGH.

(1885.)

Large 8vo, Cloth, pp. xvi., 783, Price **25s.** *Illustrated with 226 Wood Engravings, and 68 pages of Lithograph Plates, exhibiting 91 Figures—317 Illustrations in all.*

DISEASES OF THE HEART AND THORACIC AORTA.

By BYROM BRAMWELL, M.D., F.R.C.P.Ed.,

LECTURER ON THE PRINCIPLES AND PRACTICE OF MEDICINE, AND ON PRACTICAL MEDICINE AND MEDICAL DIAGNOSIS, IN THE EXTRA-ACADEMICAL SCHOOL OF MEDICINE, EDINBURGH ; ASSISTANT PHYSICIAN EDINBURGH ROYAL INFIRMARY.

(1884.)

8vo, Cloth, pp. xvi., 359, Price **16s.**, *with 183 Illustrations, including 53 pages of Lithograph Plates printed in Colours.*

SECOND EDITION, RE-WRITTEN AND ENLARGED

DISEASES OF THE SPINAL CORD.

By BYROM BRAMWELL, M.D., F.RC.P.Ed.,

LECTURER ON THE PRINCIPLES AND PRACTICE OF MEDICINE, AND ON PRACTICAL MEDICINE AND MEDICAL DIAGNOSIS, IN THE EXTRA-ACADEMICAL SCHOOL OF MEDICINE, EDINBURGH ; ASSISTANT PHYSICIAN, EDINBURGH ROYAL INFIRMARY.

(1884.)

Large 8vo, Cloth, pp. 150, *with* 41 *Illustrations, Price* **4s. 6d.**

PRACTICAL MEDICINE AND MEDICAL DIAGNOSIS.

METHODS OF DIAGNOSIS—CASE-TAKING AND CASE RECORDING—MEDICAL THERMOMETRY.

By BYROM BRAMWELL, M.D., F.R.C.P.Ed.,

LECTURER ON THE PRINCIPLES AND PRACTICE OF MEDICINE, AND ON PRACTICAL MEDICINE AND MEDICAL DIAGNOSIS, IN THE EXTRA-ACADEMICAL SCHOOL OF MEDICINE, EDINBURGH ; ASSISTANT PHYSICIAN, EDINBURGH ROYAL INFIRMARY.

(1887.)

Foolscap 8vo, Cloth, pp. viii., 153, *Price* **3s. 6d.**

SYNOPSIS OF CHEMISTRY,

INORGANIC AND ORGANIC,

TO ASSIST STUDENTS PREPARING FOR EXAMINATIONS.

By THOS. W. DRINKWATER, F.C.S.,

LECTURER ON CHEMISTRY IN THE EDINBURGH SCHOOL OF MEDICINE.

(1882.)

Crown 8vo, pp. 233, *with* 117 *Illustrations, Fourth Edition, Revised and Enlarged, Price* 4s. 6d.

A COMPEND OF ANATOMY,
INCLUDING VISCERAL ANATOMY.
By SAM^L. O. L. POTTER, M.A., M.D.,
COOPER MEDICAL COLLEGE, SAN FRANCISCO.

(1887.)

Crown 8vo, pp. 316, *Second Edition, Revised and Enlarged, Price* 7s. 6d.

A COMPEND OF THE PRACTICE OF MEDICINE.
By DANIEL E. HUGHES, M.D.,
DEMONSTRATOR OF CLINICAL MEDICINE IN THE JEFFERSON MEDICAL COLLEGE OF PHILADELPHIA.

(1886.)

Crown 8vo, pp. 120, *Second Edition, Revised and Enlarged, with* 22 *Illustrations, Price* 4s. 6d.

A COMPEND OF OBSTETRICS.
By HENRY G. LANDIS, A.M, M.D.,
PROFESSOR OF OBSTETRICS AND DISEASES OF WOMEN IN STARLING MEDICAL COLLEGE.

(1886.)

Crown 8vo, pp. 156, *Second Edition, Revised and Enlarged, with* 62 *Illustrations, Price* 4s. 6d.

A COMPEND OF SURGERY,
FOR STUDENTS AND PHYSICIANS.
By ORVILLE HORWITZ, B.S., M.D.,
RESIDENT PHYSICIAN AT THE PENNSYLVANIA HOSPITAL.

(1886.)

8vo, Cloth, pp. 172, *with* 16 *Wood Engravings, Price* 7s. 6d.

A PRACTICAL TREATISE ON
IMPOTENCE, STERILITY,
AND ALLIED DISORDERS OF THE MALE SEXUAL ORGANS.
By SAMUEL W. GROSS, A.M., M.D., LL.D.,
PROFESSOR OF THE PRINCIPLES OF SURGERY AND CLINICAL SURGERY IN THE JEFFERSON MEDICAL COLLEGE OF PHILADELPHIA.

(1887.)

8vo, Cloth, pp. xv., 283, with 42 Wood Engravings, Price 9s.

THE DISEASES OF THE EAR AND THEIR TREATMENT.

By ARTHUR HARTMANN, M.D., BERLIN.

TRANSLATED FROM THE THIRD GERMAN EDITION BY

JAMES ERSKINE, M.A., M.B.,

SURGEON FOR DISEASES OF THE EAR TO ANDERSON'S COLLEGE DISPENSARY, GLASGOW; LATE ASSISTANT-SURGEON TO THE GLASGOW HOSPITAL AND DISPENSARY FOR DISEASES OF THE EAR.

(1887.)

Large 8vo, pp. xiv., 600, with 147 Illustrations, some coloured, Price 30s.

THE REFRACTION AND ACCOMMODATION OF THE EYE

AND THEIR ANOMALIES.

By E. LANDOLT, M.D.,

PROFESSOR OF OPHTHALMOLOGY, PARIS.

TRANSLATED UNDER THE AUTHOR'S SUPERVISION BY

C. M. CULVER, M.A., M.D.,

FORMERLY CLINICAL ASSISTANT TO THE AUTHOR; MEMBER OF THE ALBANY INSTITUTE, ALBANY, N.Y.

(1886.)

Large 8vo, pp. xxviii., 772, with 404 Illustrations, Price 31s. 6d.

THE PARASITES OF MAN:

AND THE DISEASES WHICH PROCEED FROM THEM.

A TEXT-BOOK FOR STUDENTS AND PRACTITIONERS.

By RUDOLF LEUCKART,

PROFESSOR OF ZOOLOGY AND COMPARATIVE ANATOMY IN THE UNIVERSITY OF LEIPSIC.

Translated from the German, with the Co-operation of the Author,

By WILLIAM E. HOYLE, M.A. (OXON.), M.R.C.S., F.R.S.E.

NATURAL HISTORY OF PARASITES IN GENERAL. SYSTEMATIC ACCOUNT OF THE PARASITES INFESTING MAN. PROTOZOA—CESTODA.

(1886.)

Royal 4to, Extra Cloth, Price 63s. *nett.*

ATLAS OF VENEREAL DISEASES.

A Series of Illustrations from Original Paintings, with Descriptions of the varied Lesions, their differential Diagnosis and Treatment.

By P. H. MACLAREN, M.D., F.R.C.S.E.,

SURGEON, EDINBURGH ROYAL INFIRMARY; FORMERLY SURGEON IN CHARGE OF THE LOCK WARDS, EDINBURGH ROYAL INFIRMARY; EXAMINER IN THE ROYAL COLLEGE OF SURGEONS, EDINBURGH.

(1887.)

Large 8vo, Cloth, pp. 701, *with* 214 *Wood Engravings, and a Coloured Plate, Price* 18s.

THE SCIENCE AND ART OF OBSTETRICS.

By THEOPHILUS PARVIN, M.D., LL.D.,

PROFESSOR OF OBSTETRICS AND DISEASES OF WOMEN AND CHILDREN IN JEFFERSON MEDICAL COLLEGE, PHILADELPHIA, AND ONE OF THE OBSTETRICIANS TO THE PHILADELPHIA HOSPITAL.

(1887.)

Crown 8vo, Cloth, Price 4s. 6d.

THE NURSING AND CARE OF THE NERVOUS AND THE INSANE.

By CHARLES K. MILLS, M.D.,

PROFESSOR OF DISEASES OF THE MIND AND NERVOUS SYSTEM IN THE PHILADELPHIA POLYCLINIC AND COLLEGE FOR GRADUATES IN MEDICINE; LECTURER ON MENTAL DISEASES IN THE UNIVERSITY OF PENNSYLVANIA.

(1887.)

Crown 8vo, Cloth, Price 4s. 6d.

MATERNITY, INFANCY, CHILDHOOD.

HYGIENE OF PREGNANCY; NURSING AND WEANING OF INFANTS; THE CARE OF CHILDREN IN HEALTH AND DISEASE.

Adapted especially to the use of Mothers or those intrusted with the bringing up of Infants and Children, and Training Schools for Nurses, as an aid to the teaching of the Nursing of Women and Children.

By JOHN M. KEATING, M.D.,

LECTURER ON THE DISEASES OF WOMEN AND CHILDREN, PHILADELPHIA HOSPITAL.

(1887.)

Crown 8vo, Cloth, Price **4s. 6d.**

OUTLINES FOR THE MANAGEMENT OF DIET:

Or, THE REGULATION OF FOOD TO THE REQUIREMENTS OF HEALTH AND THE TREATMENT OF DISEASE.

By E. M. BRUEN, M.D.

(1887.)

Crown 8vo, Cloth, Price **4s. 6d.**

FEVER NURSING:

INCLUDING

1. On Fever Nursing in General. 2. Scarlet Fever. 4. Enteric or Typhoid Fever. 4. Pneumonia and Rheumatism.

By J. C. WILSON, M.D.

(1887.)

8vo, Cloth, pp. xii., 302, *with* 5 *Wood Engravings, Price* **9s.**

DISEASES OF THE MOUTH, THROAT, AND NOSE.

INCLUDING

RHINOSCOPY AND METHODS OF LOCAL TREATMENT.

By PHILIP SCHECH, M.D.,
LECTURER IN THE UNIVERSITY OF MUNICH.

TRANSLATED BY R. H. BLAIKIE, M.D., F.R.S.E.,
FORMERLY SURGEON, EDINBURGH EAR AND THROAT DISPENSARY; LATE CLINICAL ASSISTANT, EAR AND THROAT DEPARTMENT, ROYAL INFIRMARY, EDINBURGH.

(1886.)

8vo, Cloth, pp. xii., 223, *with* 7 *Illustrations, Price* **9s.**

ELEMENTS OF PHARMACOLOGY.

By DR. OSWALD SCHMIEDEBERG,
PROFESSOR OF PHARMACOLOGY, AND DIRECTOR OF THE PHARMACOLOGICAL INSTITUTE, UNIVERSITY OF STRASSBURG.

TRANSLATED UNDER THE AUTHOR'S SUPERVISION

By THOMAS DIXSON, M.B.,
LECTURER ON MATERIA MEDICA IN THE UNIVERSITY OF SYDNEY, N.S.W.

(1887.)

New and thoroughly Revised Edition, large 8vo, Cloth, pp. 877, Illustrated with over 1000 Wood Engravings. Price 24s.

THE PRINCIPLES AND PRACTICE OF OPERATIVE SURGERY.

By STEPHEN SMITH, A.M., M.D.,

PROFESSOR OF CLINICAL SURGERY IN THE UNIVERSITY OF THE CITY OF NEW YORK; SURGEON TO BELLEVUE AND ST. VINCENT HOSPITALS, NEW YORK.

(1887.)

8vo, Cloth, pp. 385, *with Illustrations, Price* 12s. 6d.

DISEASES OF THE DIGESTIVE ORGANS IN INFANCY AND CHILDHOOD,

WITH CHAPTERS ON THE INVESTIGATION OF DISEASE AND ON THE GENERAL MANAGEMENT OF CHILDREN.

By LOUIS STARR, M.D.,

CLINICAL PROFESSOR OF DISEASES OF CHILDREN IN THE HOSPITAL OF THE UNIVERSITY OF PENNSYLVANIA; PHYSICIAN TO THE CHILDREN'S HOSPITAL, PHILADELPHIA.

(1886.)

18mo, *Cloth, pp.* iv., 150, *Price* 5s. 6d.

GERMAN-ENGLISH MEDICAL DICTIONARY.

By JOSEPH R. WALLER, M.D.

(1885.)

Large 8vo, Cloth, pp. xx., 762, *with upwards of* 250 *Engravings, and two full-page Plates, Price* 25s.

A TREATISE ON AMPUTATIONS OF THE EXTREMITIES,

AND THEIR COMPLICATIONS.

By B. A. WATSON, M.D.,

SURGEON TO THE JERSEY CITY CHARITY HOSPITAL, TO ST. FRANCIS' AND TO CHRIST'S HOSPITAL AT JERSEY CITY, N.Y.

(1885.)

Crown 8vo, pp. 674, *with* 117 *Illustrations, Price* 15s.

DISEASES OF WOMEN.
A HANDBOOK FOR PHYSICIANS AND STUDENTS,
By Dr. F. WINCKEL,
PROFESSOR OF GYNÆCOLOGY, AND DIRECTOR OF THE ROYAL UNIVERSITY CLINIC FOR WOMEN, IN MUNICH.

AUTHORISED TRANSLATION BY
J. H. WILLIAMSON, M.D.,
RESIDENT PHYSICIAN GENERAL HOSPITAL, ALLEGHENY, PENNSYLVANIA.

UNDER THE SUPERVISION, AND WITH AN INTRODUCTION BY
THEOPHILUS PARVIN, M.D.,
PROFESSOR OF OBSTETRICS AND DISEASES OF WOMEN AND CHILDREN IN JEFFERSON MEDICAL COLLEGE, PHILADELPHIA; AUTHOR OF "THE SCIENCE AND ART OF OBSTETRICS."

(1887.)

8vo, Cloth, pp. xvi., 534, *Illustrated with* 162 *Coloured Plates, mostly from Original Drawings, Price* 24s.

SECOND EDITION, REVISED AND IN PART RE-WRITTEN.
PRACTICAL PATHOLOGY:
A MANUAL FOR STUDENTS AND PRACTITIONERS,
By G. SIMS WOODHEAD, M.D., F.R.C.P.Ed.,
FORMERLY DEMONSTRATOR OF PRACTICAL PATHOLOGY IN THE UNIVERSITY OF EDINBURGH; PATHOLOGIST TO THE ROYAL INFIRMARY, EDINBURGH.

(1885.)

8vo, Cloth, pp. xii., 174, *with* 60 *Illustrations, mostly Original* (34 *in Colours*), *Price* 8s. 6d.

PATHOLOGICAL MYCOLOGY:
AN INQUIRY INTO THE ETIOLOGY OF INFECTIVE DISEASES.
By G. SIMS WOODHEAD, M.D., F.R.C.P.Ed.,
PATHOLOGIST TO THE ROYAL INFIRMARY, EDINBURGH,

AND

ARTHUR W. HARE, M.B., C.M.,
ASSISTANT TO THE PROFESSOR OF SURGERY IN THE UNIVERSITY OF EDINBURGH.

SECTION I.—METHODS.
(1885.)

YOUNG J. PENTLAND, EDINBURGH.

www.ingramcontent.com/pod-product-compliance
Lightning Source LLC
Chambersburg PA
CBHW030335170426
43202CB00010B/1131